Preface to the First Edition

This HANDBOOK has been written to supply in concise form much of the information required by the new Advanced Level syllabuses in biology. A number of examining boards have made extensive revisions in these syllabuses during the last three years to bring them more in line with current trends in the subject. With the permission of the Associated Examining Board, the Oxford and Cambridge Schools Examinations Board, the Cambridge Local Syndicate and the University of London Board, questions based on the new syllabuses are included at the end of appropriate chapters.

All the syllabuses considered have a common corpus of knowledge about the cell and its functioning with which students should be familiar. They also have a similar content on genetics and the theory of evolution. The rest of the syllabuses are, in the main, made up of physiological functions studied from a biological standpoint, rather than from separate zoological and botanical approaches. The flowering plant and mammal are the major examples considered but for the remainder the old convention of "set types" has been dispensed with and wide selection of comparative systems can be made. While this makes for a much wider approach to the subject, it does present certain difficulties for the candidate and the questions that are set are necessarily of the "open-ended" type, which allow free choice of material.

The method employed in this HANDBOOK is to deal with the common material of the new syllabuses and then to select examples of comparative functions from a wide variety of organisms. The first chapters deal with the classification of living things so that the systematic position of the examples used can be determined. The actual systems and organisms described are not necessarily the best or the most apposite since an element of free choice should be exercised throughout.

In Appendix II the method of answering "A" Level biology questions is described and in Appendix III there are examples of the multi-choice type of question that have been set recently by examining boards. The author would like to emphasise that the references made to works in the Bibliography should be followed up for a study in depth to be made. Also, biology is essentially a practical and experimental subject and while these aspects have no place in a concise HANDBOOK it is to be hoped that candidates will approach the subject from an investigative point of view.

Note on the capitalisation of names. *All* Latin names, e.g. Bryophyta, Reptilia, must have an initial capital letter. Certain of these names have English equivalents, e.g. bryophyte, reptile, in which case no initial capital is required. However, as only certain names have English equivalents, in headings and when the word is first introduced, the Latin form is used. Generally, the English version is used in the text, although it is important that both forms are found, in order to emphasise to the student that two forms exist.

In certain cases, the Latin version has become the accepted English form, and so Algae and algae are to be found; another example is Fungi and fungi.

Note on SI units. SI units are used in this text and a table of those units that the biology student will encounter most commonly is set out below:

Physical quantity	SI unit	Symbol
Length	metre	m
Mass	kilogram	kg
Time	second	s
Electric current	ampere	A
Temperature	degree Celsius (common)	°C
	kelvin (absolute)	K
Energy	joule	J

Some useful conversion factors are as follows:

$$\mu m = 10^{-6} \text{ m}$$
$$1 \text{ litre} = 10^{-3} \text{ m}^3$$
$$1 \text{ cc} = 10^{-5} \text{ m}^3 = 1 \text{ ml (approx.)}$$
$$1 \text{ calorie} = 4 \cdot 2 \text{ J}$$
$$760 \text{ mm Hg (1 atmosphere)} = 101 \text{ kPa}$$

Acknowledgments. The author would like to express his thanks to the following for permission to reproduce or to modify their material:

The Syndics of the Cambridge University Press for Figs. 37,

78, 102, 120 and 153 from *Physiology of Mammals and other Vertebrates* by P. T. Marshall and G. M. Hughes.

W. B. Saunders & Company for Fig. 13 from *The Fundamentals of Microbiology* by M. Frobisher.

Longmans for Fig. 40, amended from *Guide to Subcellular Botany* by C. A. Stace.

Blackie & Son for Fig. 20 from *Floral Anatomy* by S. G. Jones.

Macmillan for Fig. 50 amended from *Elementary Genetics* by Wilma George (1965 edition).

The Oxford University Press for Fig. 55 taken from *Plant Physiology* by W. O. James.

The Editor of the *School Science Review* for Fig. 64 from "Audus" (1953, vol. 125).

Sir Isaac Pitman & Sons for Fig. 70 from *Plant and Animal Biology* by A. E. Vines and N. Rees.

John Murray for Fig. 99 from *Biological Drawings* by Maud Jepson.

Heinemann for Fig. 101 amended from *Histology* by W. H. Freeman and B. Bracegirdle.

Prentice-Hall for Fig. 126 amended from *Animal Behaviour* by V. G. Dethier and E. Stellar.

The Pergamon Press for Fig. 162 from *The Development of Modern Biology* by P. T. Marshall.

The author would also like to thank Mr Peter Redmond and the editorial staff of Macdonald and Evans for their help and encouragement with the manuscript during its preparation.

The book was read at proof stage by Mr A. I. Reiff, Senior Science Master at Taft School, Connecticut, U.S.A. while on exchange at The Leys. I am most grateful for his advice and corrections.

Most of this book was written during a sabbatical term spent at Christ Church, Oxford, and at the Zoological Department of the University. I would like to acknowledge the help, advice and facilities for work made available to me while in Oxford.

P. T. M.

February 1970

Preface to the Third Edition

In the major revision of this HANDBOOK, carried out for this latest edition, all examination questions have been renewed and with the permission of the relevant boards the new questions included at the ends of chapters are all from very recent years. It is now the policy of most boards to detail marks and sub-marks on examination papers, and this information will be found of very considerable help in tackling problems.

As far as the text is concerned an effort has been made in all areas to incorporate recent findings and changes of emphasis. Textbooks naturally lag behind research discoveries but by keeping a careful watch on monographs and publications it is possible to detect subtle changes in emphasis. Often topics become clearer as we know more about them, but sometimes the process works in reverse and what appeared to be straightforward is now seen to be extremely complex.

Readers will find the sections on photosynthesis, respiration, transpiration, hormones and genes, and some aspects of the nervous system have been taken a little further than in previous editions. In the section on evolution a more experimental approach to natural selection is included.

It appears that during the eight years this HANDBOOK has been in print it has been, as intended, a helpful back up to lecture notes and in revision for "A" level examinations in Biology. The author is grateful to those students who have either communicated directly with him or written to the publishers, with helpful comments and suggestions. As they will see, many of these have been incorporated into this new edition.

P. T. M.

August 1978

Contents

PART ONE

THE PRINCIPLES OF BIOLOGY

CHAPTER 1

The Basis of Classification

1. Introduction. As already mentioned in the Preface, the new biology syllabuses have moved away from the idea of the "set type" of example, so dear to classical studies. These set types were taken as representatives of important biological groups or levels of organisation and the student needed to know their morphology and functioning in considerable detail. The changing emphasis towards biology as an experimental rather than a descriptive science has led to much wider ideas about the variety of organisms that should be included in a modern introductory course at the sixth-form level. It is useful for the student to keep this principle in mind as he works through the course. It is hoped that the student will make himself familiar with a wide variety of organisms so that appropriate examples can be selected to illustrate biological principles. This process of selection should be exercised by both the teacher and student and should allow considerable freedom of choice.

While it is the intention of the various boards to encourage such an individual approach, according to the material that might be locally available, recommend itself, or be more familiar, there does remain a framework of biological classification that should be known. Such a disciplined classification should be looked upon as a form of vocabulary from which meaningful sentences, or ideas, can be constructed. It is also a common part of all biological courses that students should have a broad knowledge of the plant and animal kingdoms and be able to appoint to its correct phylum or group any reasonably well-known living organism that they come across.

In this first chapter the evolutionary basis and outlines of biological classification will be presented and while it is recommended that students should generally familiarise themselves with this "who's who" before proceeding further, the following points should be borne in mind:

(*a*) The catalogue of organisms serves as a reference which may be consulted throughout the course.

(*b*) A real knowledge of biological groups can be achieved only by examination of actual specimens.

(*c*) The following classification is only an outline and necessarily selective. Other examples can be used quite as well for functional studies and are equally acceptable for examination answers.

SYSTEMS OF CLASSIFICATION

2. The binomial system of classification. All classification, biological or otherwise, is only a human activity that allows ease of description and recognition. Even before the unifying principle of organic evolution was recognised valuable classifications were constructed dating back to Aristotle in 384 B.C. The most useful classification, and that used in a modified form today, was that of Linnaeus which was propounded in the tenth edition of his *Systema Natura* (1758). This method is called the *binomial system* and each organism is given two names; the *specific* name, defining the particular type of organism, and the *generic* name, which it may share with other closely-related forms.

Closely-related genera are put together in *families* and a number of families make up an *order*, the differences at each level growing more complete. A *class* is made up of a number of *orders* and is itself the sub-unit of the very large natural groupings that are called *phyla*. Finally, the various phyla come into one of the two great *kingdoms* of living organisms—the *plants* and the *animals*.

NOTE: For very specialised classifications the use of numerous further divisions of the above scheme is used, e.g. between an order and a class there may be a super-order and a sub-class. Such detail, however, is mostly outside the scope of "A" Level work.

3. Exercise in classification: the classification of man. A reasonably comprehensive classification of man on the above system would be as follows:

Kingdom: Animal (*see* I, **6**).

Phylum: Chordata (*see* II, **9**).

Sub-phylum: Vertebrata (or Craniata) (*see* II, **10**).

Class: Mammalia (*see* II, **10**(*f*)).

Sub-class: Eutheria (or Placentalia). The young are born alive and develop in the uterus of the female: they are attached to this by the placenta. Mammary glands are present.

Order: Primates. There is extreme cerebral development, five digits, one of which is opposable, the radius and ulna are free to rotate. They are mostly arboreal.

Sub-order: Anthropoidae. This includes monkeys, apes and men. They are arboreal or secondarily terrestrial, the face is hairless; cerebral development is very marked and usually there is only one offspring at birth.

Super-family: Hominoidea.

Family: Hominidae.

Genus: Homo. He has an upright gait and a cranial capacity exceeding 1100 ml.

Species: sapiens. There is a small muscular ridge development on the cranium and above the eyes; jaws and teeth are small, and there is a well-developed chin.

Man therefore is classified as *Homo sapiens* or *H. sapiens L.*, the generic name having a capital letter, but not the specific and, for a fully correct form, the initial letter or abbreviation of the original classifier (in this case Linnaeus) is included.

4. Ideas implicit in biological classifications. It is possible to build up ordered schemes of classification in biology because the process of evolution means that one group has diverged from another. A *natural classification* is one based on evolutionary affinities rather than on superficial similarities.

If we classified together all flying animals we should find bats, birds and insects all in the same group. Such a classification would have taken only one characteristic into account and is an example of an artificial, i.e. a non-evolutionary, scheme. The classing of flowers by their colour would be a similar example.

Sometimes it is very difficult to be sure of the evolutionary affinities of naturally occurring groups but as far as may be possible biologists aim at natural rather than artificial classifications.

MAJOR DIVISIONS OF LIVING ORGANISMS

5. Characteristics of living organisms. Living organisms are commonly separated from non-living organisms by the following features:

(*a*) *Nutrition:* the taking in of substances that can provide metabolic energy and materials for growth.

(*b*) *Respiration:* the breaking down of substrates for the provision of energy.

(*c*) *Excretion:* the passing out of the wastes of metabolism.

(*d*) *Growth:* the synthesis of new material within the organism leading to increase in size.

(*e*) *Reproduction:* the means whereby the individual organism gives rise to another.

(*f*) *Irritability:* the response of the organism to changes in both internal and external conditions.

(*g*) *Movement:* the means by which the whole organism or a part of it is able to move (a feature which is much more evident in animals than plants).

NOTE: While these characteristics generally apply, a group of disease-causing organisms called viruses have no means of respiration. They may also be crystallised and remain inert for indefinite periods. It is thus not easy to classify them as living or non-living. A recent idea was to include the possession of nucleic acid as one of the essential characteristics of living things. This would include viruses as living organisms. (*See also* III, **13–18**.)

6. Defining the kingdoms. Living organisms fall into one of two kingdoms, which may be defined as follows:

(*a*) Plants feed *holophytically* (*see* V1, **1**) and are able to manufacture their food from carbon dioxide, water and inorganic ions in the presence of chlorophyll and light. The process is called *photosynthesis*. Because of these nutritional requirements plants tend to be sessile, rigid objects whose cell walls are supported by the inert polysaccharide cellulose. Plant cells also have large fluid-filled vacuoles. The storage substance of most plants is starch.

NOTE: Fungi were probably derived from green plants but have lost the ability to photosynthesise and do not possess chlorophyll. They require an organic source of food and tend to take this in soluble form all over their surfaces. This means of nutrition is called *saprophytic* and allows very rapid growth but has the disadvantage of restricting the plant to living in damp places or in liquids.

Some fungi have cell walls of chitin and storage material of glycogen. Their cells have large vacuoles and they tend to grow in long threads called *hyphae*.

(*b*) Animals need an organic source of food and take this into their bodies by means of a mouth; this form of feeding is called *holozoic*. Because they feed on other animals or green plants, they need to be motile and only a few filter and particle feeders are sessile. Animals have no rigid cell walls and their cells do not contain large vacuoles. Glycogen is commonly used as a storage substance.

NOTE: It must again be emphasised that these criteria are by no means absolute and that many plants, especially among the green algae, have powers of locomotion. There are also motile fungi and fungal spores. As far as nutrition is concerned the divisions are fairly sharp, although, as is well known, certain insectivorous plants feed on animals for their source of nitrogen.

PROGRESS TEST 1

1. What is meant by the "binomial system of classification"? (2)

2. To what order and what class of what phylum does man belong? (3)

3. What are the characteristics of living organisms? (5)

4. How can animals and plants be distinguished as two separate groups? (6)

Examination questions on classification are to be found at the end of Chapter III.

CHAPTER II

Animal Classification

SINGLE-CELLED ANIMALS

1. Protozoa. Protozoa are *single celled* and mainly microscopic. The more primitive members have certain plant characteristics and at this level the distinction between animals and plants breaks down. (The term Protista is also used for unicellular organisms.)

The structural systems within the protozoan cell are termed *organelles* and the group is very diverse in its morphology and physiology. It may have arisen from several different stocks of plants. There are some 50,000 species of Protozoa distinguished by biologists and the phylum has been sub-divided in a number of ways. A scheme which follows some contemporary ideas is given below:

(*a*) *Sub-phylum Sarcomastigophora* have pseudopodia or flagella or both. They have only one nucleus and reproduce by fission. The main classes are the Phytomastigophora which have plant-like characters and normally feed holophytically. They include certain members of the phytoplankton as well as the species Euglena (*see* Fig. 1).

Contrasting with the above are the class Zoomastigophora which are flagellated forms and always feed holozoically or parasitically. These include the "collar-flagellates" which show some affinities with the sponges, as well as the Trypanosoma which cause the disease sleeping sickness. An interesting form is Trichonympha which live symbiotically in the gut of the termite and are able to digest wood.

The class Rhizopoda are typically amoeboid and include Amoeba (*see* Fig. 1) as well as the "slime fungi" whose exact systematic position is somewhat obscure. They also include the Foraminifera which are important members of the plankton and whose shells accumulate to great depths on the sea bed. Geologists use Foraminifera fossils to date rock strata.

(*b*) *Sub-phylum Sporozoa* are made up of parasitic protozoans and at one stage or another in their life histories they form

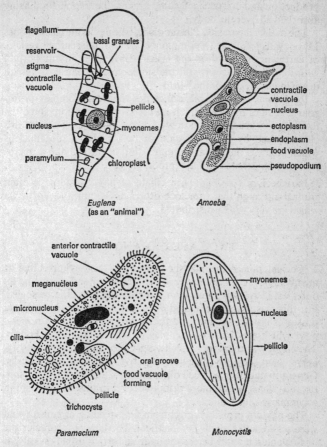

FIG. 1 *A selection of Protozoa showing some of the common organelles found among this group.*

Euglena may be considered also as a member of the green algae although in the dark certain species lose their chlorophyll and feed saprophytically.

resistant spores. They have both sexual and asexual means of re-production and no specific feeding organs, the soluble food being absorbed all over the body.

The malarial parasite, Plasmodium, belongs in the class as does Monocystis (*see* Fig. 1), commonly found in the seminal vesicles of earthworms. Whereas the latter seems to be in equilibrium with its host, malaria is still the single greatest cause of human death in many parts of the world.

(*c*) *Sub-phylum Ciliophora* includes the most complex proto-zoans. Besides the cilia, which are used for locomotion, and often feeding, they also have a variety of other organelles. They are also characterised by the possession of two nuclei, a meganucleus which controls the somatic organisation of the cell and a micro-nucleus important in sexual reproduction.

Paramecium (*see* Fig. 1), a ciliate of hay infusions, is often studied although there are such others as Stentor and Vorticella which are very widely distributed.

TWO-LAYERED ANIMALS

2. Coelenterata. A distinguished feature of this phylum is that its members have *only two layers of cells*, an inner endoderm and an outer ectoderm separated by a structureless jelly, the mesoglea. For this reason they are called *diploblastic* animals. Coelenterata possesses unique stinging and food-catching cells called *nemato-cysts* which are the most complex of all cell organelles known. They do have other types of cells, e.g. epitheliomuscular, sensory, nerve, neurosecretory, flagellate-digestive, glandular and sex cells. Generally, the degree of cell differentiation and organisation into tissue systems is at a more primitive level in this animal than in other many-celled animals.

The phylum probably arose from filter-feeding organisms that were sessile and radially symmetrical. The latter is characteristic of Coelenterata but whereas the hydroid form is usually sessile, as in the sea anemones, the medusoid form is motile.

(*a*) *Class Hydrozoa* includes those coelenterates which norm-ally have both hydroid and medusoid forms present in the life history. Typical are Obelia or Campanularia (*see* Fig. 2), found in colonies growing on the seaweed of rocky shores. The hydroid form is branched and grows out by asexual budding occasionally giving rise to special medusae-producing regions. Free-swimming

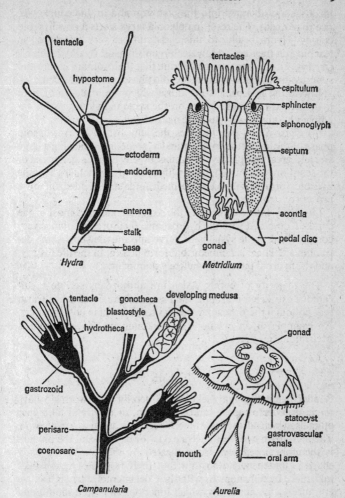

FIG. 2 *Different forms of Coelenterata.*
The medusoid form may be considered as an inverted hydroid.

medusae are liberated into the plankton and in due course give rise to sex cells. The resulting planula larva starts a new hydroid colony. Hydra belong to this class and are very often used in teaching as they are freshwater organisms and readily available (*see* Fig. 2). It is worth remembering that their solitary nature and suppression of a free-living medusoid phase are very atypical.

Interesting hydrozoans are those which are made up of both hydroid and medusoid forms on the same individual, as for example Physalia (the Portuguese man-of-war).

(*b*) *Class Scyphozoa* includes the jelly fishes. The medusoid phase of the life history predominates, sometimes reaching a large size and great complexity. A number of jelly fishes, e.g. Aurelia of the sea shore (*see* Fig. 2), have a life cycle which includes a sessile hydroid form from which the medusoids are budded off by a process of strobilisation.

(*c*) *Class Anthozoa* includes the sea anemones and most of the corals. Compared with the hydrozoans, their organisation is very complex and tissue systems of nerve and muscle show some development. There is no medusoid form present in the life history, and the hydroid phase is completely dominant.

The majority of corals which are anthozoans secrete a calcareous exoskeleton around the base, and are confined to warm seas although there are a few species to be found in colder waters. An example of an anemone to be found on the sea shore is Metridium (*see* Fig. 2).

WORMS

3. Platyhelmintha. Platyhelmintha (literally "flat worms") have *three layers* ecto-, meso- and endoderm, so they, and all higher animals which also have three layers, are termed *triploblastic*. The mesoderm does not have a cavity or coelom and the phylum is therefore *acoelomate*. As indicated by their name the body shape is flattened and so thin that their respiratory exchanges take place by diffusion. As with the Coelenterata, there is a single entrance to the gut cavity which functions as both a mouth and an anus. Excretion and osmoregulation are carried out by characteristic flame-cells. The majority of flat worms are hermaphrodite and correlated with this, they usually have complicated sex organs and various means of preventing self-fertilisation.

Of the three classes two are important as parasites of higher organisms including man and his domestic animals. The three classes are as follows:

(a) *Class Turbellaria* comprises the free-living flatworms commonly found in fresh water and the sea. *Planaria* form good examples of the former and may be caught from ditches and other similar habitats (*see* Fig. 3). Turbellarian worms have cilia covering their bodies and a fairly well-developed range of sense organs. They are active carnivores and scavengers. Biologically they are of interest as a contrast to the closely related but very highly adapted parasitic classes, as well as being suitable material for studies of grafting and regeneration.

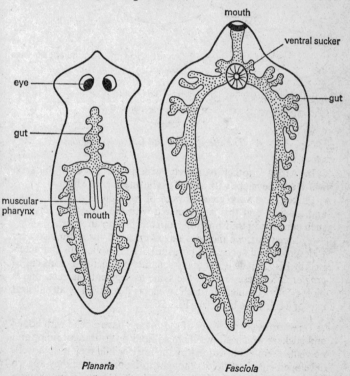

FIG. 3 *A free-living planarian worm contrasted with a liver fluke.* The former has much more elaborate sense organs.

(b) *Class Trematoda* are known as flukes, e.g. sheep liver fluke, Fasciola (*see* Fig. 3), and are found as both ecto- and endo-parasites. The epidermis of the adult secretes a cuticle but there are larval forms which resemble the free-living turbellarians. A complex life history may be found both in terms of the different larval stages as well as in the transmission from one vector to another. Such a life history is outlined in Fig. 4. (*See* Fig. 43 for greater detail.)

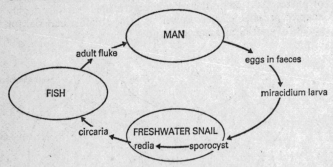

FIG. 4   *The life cycle of the Chinese liver fluke.*

In Africa another extremely widespread fluke is Bilharzia which also develops with a snail as the intermediate host.

Flukes have very great powers of reproduction, both sexual and asexual, and this is associated with the losses sustained by such parasites in transmission from one host to another.

(c) *Class Cestoda* includes tapeworms and these are also very widespread and successful parasites. Their bodies are characterised by being made up of many separate but identical units called *proglottides*: these are not the same as segments as each one is biologically complete in itself, whereas a segment is an integral part of a complete body (*see* Fig. 5).

Tapeworms are parasites of the gut and have a head with hooks and suckers which allows them to hold on to the gut lining and prevents them being voided by the host. While they have a well-developed cuticle which is thought to be protective, they absorb soluble food through this into the body. They do not have a mouth, or a gut and their muscle systems are very reduced. Some of their adaptations, such as anti-enzymes to prevent their diges-

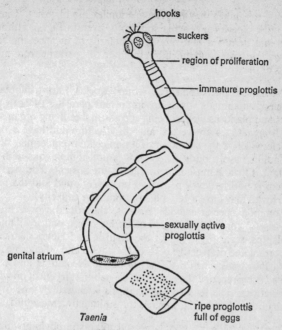

hooks

suckers

region of proliferation

immature proglottis

sexually active proglottis

genital atrium

ripe proglottis full of eggs

Taenia

FIG. 5 Taenia, *the tapeworm. These worms may be many feet in length when fully developed.*

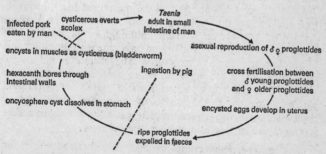

Infected pork eaten by man

cysticercus everts scolex

Taenia adult in small intestine of man

encysts in muscles as cysticercus (bladderworm)

asexual reproduction of ♂ ♀ proglottides

hexacanth bores through intestinal walls

Ingestion by pig

cross fertilisation between ♂ young proglottides and ♀ older proglottides

oncyosphere cyst dissolves in stomach

encysted eggs develop in uterus

ripe proglottides expelled in faeces

FIG. 6 *The life cycle of* Taenia solium, *the pork tapeworm.*

tion by the host and the ability to carry out anaerobic respiration, are physiological.

Like the flukes they may have complex life histories. That of
the pork tapeworm (Taenia solium) is given in Fig. 6, although
it should be noted that modern standards of hygiene and food
inspection have done much to eliminate this parasite from many
parts of the world.

4. Nematoda. Nematoda (the "round worms") are triploblastic
acoelomate animals which are important parasites of both animals and plants as well as being successful free-living organisms.

All nematodes are very similar in structure and have a thick
cuticle which is the secretion of a non-cellular epidermis. While
the adults have a body cavity, this is not homologous with the
coelom. Certain musculo-nerve cells are present. The nervous
system is poorly developed. Two sexes are found and the parasitic forms may have a very high reproductive capacity.

Common species include Ascaris, the round worm of the
pig, useful for study because of its large size and availability,
Proleptus, seen in the gut of the dogfish, and Heterodera, very
important and destructive pests of the potato and other crops.

Various hookworm diseases are due to Nematoda and widespread occurrence of these may have a debilitating effect on whole
populations. The disease Elephantiasis is caused by one of the
Nematoda, Wuchereria, which is transmitted by a mosquito.

5. Annelida. This is a very large and successful phylum which is
of interest to biologists because most other invetebrate stocks
can be related to it.

Annelida are triploblastic and the mesoderm has a cavity, a
true coelom, so the group is termed *coelomate*. The body is
divided up into similar segments, although at the head several
segments may operate together and have extensive developments
of feeding and sensory structures. The excretory and osmo-
regulatory organs are *nephridia* which are characteristic of the
phylum. The nervous system consists of paired dorsal cerebral
ganglia connected to the ganglia of the ventral nerve cord by
commissures passing around the pharynx. A thin chitinous cuticle
is present and larger chitinous rods called *chaetae*.

(a) *Class Polychaeta* comprises marine worms and are by far
the most numerous of the Annelida. They are characterised by
having large numbers of chaetae and usually by the development
of lateral outgrowth from each segment called parapodia. A
larval form called a *trochophore* (or *trochosphere*) is present (*see*

Fig. 7). The head is normally well developed with eyes and tentacles. It is possible to divide up the class according to the habitats of its members, thus the errant Polychaeta are those like Nereis, the rag worm (*see* Fig. 7), which are very active free-swimming carnivores; the tubiculous Polychaeta are those such as Myxicola which are tube-dwelling filter-feeding worms; and, finally the burrowing polychaetes are those like Arenicola (the lug worm) which are eaters of sand and mud.

All these types can easily be found on selected sea shores and again it should be emphasised that they are much more typical Annelida than the earthworm, so often studied.

(*b*) *Class Oligochaeta* include the earthworm, e.g. Lumbricus (*see* Fig. 7) and are distinguished by the small number of chaetae present. Due to the adaptations necessary for life in the soil, earthworms have less conspicuous sense and locomotory organs than the active Polychaeta: in fact, they compare much more closely with the burrowing forms. In order to reproduce successfully on land earthworms have a means of transfer of gametes and means for the nutrition and protection of the young in a cocoon. There is no trochophore larva.

Earthworms are very important animals which contribute extensively to the fertility of the soil.

(*c*) *Class Hirudinea* comprises the leeches (*see* Fig. 7), which are usually *ectoparasites*, although some of them feed as carnivores. They are freshwater animals and can be distinguished from other annelids by the suckers at each end of the body. There is no larval form and the eggs develop into small versions of the adult.

JOINT-LIMBED ANIMALS

6. Arthropoda. These comprise some eight-tenths of all animal species. One of their main features is the chitinous exoskeleton (also further strengthened by calcium carbonate in Crustacea). The Arthropoda are segmented animals although cephalisation is pronounced in the higher groups. Each segment bears a pair of appendages and these show adaptive radiation as mouthparts, limbs, mating and respiratory organs, again in the more advanced members. The coelom is vestigial and the body cavity, in which the blood flows, haemocoelic. There is a dorsal contractile heart. Arthropoda tend to have a well-developed nervous system which is like that of the Annelida, i.e. a dorsal cerebral ganglia and ventral nerve cord.

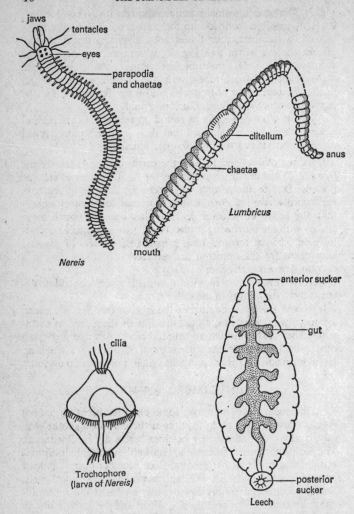

FIG. 7 *Selected annelids showing the three main classes.*
The polychaetes (represented by *Nereis*) are by far the largest and most
diverse group.

Sub-division of the phylum are as follows:

(*a*) *Class Insecta.* In this class the body is divided into three parts, a head, thorax and abdomen. There are two pairs of wings and three pairs of legs on the thorax while the abdomen carries no appendages. One pair of antennae is present. Insects respire by a tracheal system which opens to the exterior by means of segmental spiracles.

Very primitive insects are wingless, but most modern orders are winged. Of the latter, some, once again the most primitive, show incomplete metamorphosis with a nymph that resembles the adult. The wings develop outside the body. Examples of this type are the cockroach (*see* Fig. 8), locust and grasshopper of the order Orthoptera.

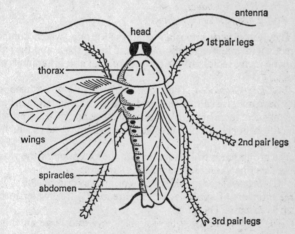

FIG. 8   *A typical insect, the cockroach.*

Because of its relatively large size this insect is often used for laboratory studies.

More advanced insects have a complete metamorphosis with complete change of form between the immature larva and the adult or imago. Metamorphosis takes place within a pupa. Examples of this type are the following:

Order Lepidoptera, e.g. butterflies, moths.
Order Coleoptera, e.g. beetles.

Order Hymenoptera, e.g. ants, bees, wasps.
Order Diptera, e.g. flies, mosquitoes.

Insects are the most successful of all animals, in terms of the number of different species that exist. They have colonised all sorts of terrestrial and freshwater environments and can live on a wide range of food substances. Many insects such as flies, mosquitoes, fleas, lice, etc. are parasites of other animals.

Features that contribute to the success of insects are as follows:

(*i*) *Adaptability of chitin* to form teeth, claws, wings and other structures.

(*ii*) Intense activity and *high metabolic rate*.

(*iii*) The ability to *fly*.

(*iv*) *High reproductive rate*.

(*v*) *Elaborate sense organs*.

They are a very important biological group, which compete directly with man for food supplies and living space and they also provide many vectors of the diseases of man and his animals and crops.

(*b*) *Class Crustacea*. The second largest class of Arthropoda, they have massive calcareous exoskeletons and the abdomen bears appendages; there are two pairs of antennae and gills are present. The life history typically involves free-swimming larvae such as the nauplius.

The group tends to be aquatic with the exception of the woodlouse (*see* Fig. 9). Sub-divisions are as follows:

(*i*) *Branchiopoda*. These are primitive crustaceans, with large numbers of appendages: they are mostly freshwater forms, such as Artemia, Daphnia.

(*ii*) *Copepoda*. They have a single eye and large antennae; they also swim. Very common examples are zooplanktons, e.g. Calanus, Cyclops (*see* Fig. 9).

(*iii*) *Cirripedia*. They are sessile, with calcareous plates in the skin; they are filter feeders, e.g. Balanus (barnacle).

(*iv*) *Malacostraca*. These are larger crustaceans with a head and thorax sometimes fused; there is a reduction of appendages. This group includes the Amphipoda such as Gammarus which are flattened from side to side, and the Isopoda with dorsiventral flattening, e.g. woodlouse. The most advanced Malacostraca are the Decapods which include crabs, lobsters and the crayfish, Astacus (*see* Fig. 9).

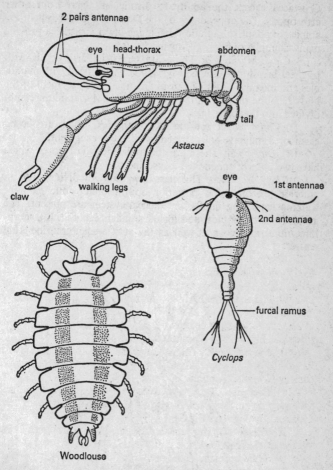

FIG. 9 *Various types of Crustacea.*

The decapods, illustrated by *Astacus*, but also including lobsters and crabs, are the most advanced group.

Little classificatory detail is required to be known about the Crustacea except the ability to distinguish them from other arthropods. The appendages of the Decapoda give excellent examples of adaptive radiation, while the lower Crustacea provide many examples of filter-feeding mechanisms.

(c) *Class Myriapoda.* These are terrestrial arthropods with tracheal breathing systems. Myriapoda have a head with mandibles and maxillae and one pair of antennae. Each segment of the body has one or two pairs of appendages. Myriapoda are subdivided into the following:

(i) *Chilopoda.* These are the centipedes, which are jawed carnivores and have one pair of appendages per segment.

(ii) *Diplopoda.* These are the millipedes, herbivores which have two pairs of appendages per segment.

(d) *Class Arachnida.* This class have the body divided into an anterior prosoma and posterior opisthosoma. Four pairs of walking legs, lung books, or tracheal system are present. The group feeds on liquids, and includes spiders (*see* Fig. 10), scorpions, mites and ticks as well as the very ancient marine king crabs.

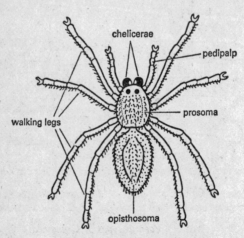

FIG. 10 *An arachnid spider.*

Scorpions and mites are also members of the Arachnida.

LESS IMPORTANT INVERTEBRATE PHYLA

7. Mollusca. This phylum is not required to be known in any detail but individual species do furnish excellent examples of certain biological principles, e.g. herbivorous adaptation in the snails, filter-feeding mechanisms in bivalves, brain and behaviour in cephalopoda, hermaphrodite reproductive mechanisms in snails, etc.

The phylum may be distinguished by the possession of a foot, a visceral mass, a mantle, which secretes the shell and a mantle cavity containing gills which in terrestrial forms acts as a lung. Mollusca are divided into three main classes:

(*a*) *Gastropoda* usually have coiled shells and a well-developed head, e.g. the snail Helix (*see* Fig. 11), the slug, the whelk.

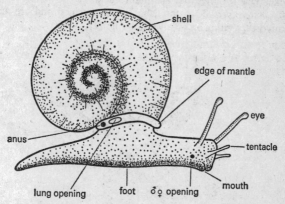

FIG. 11 *The snail is a gastropod mollusc and these are the only types to have colonised land.*

Slugs are close relatives of snails in which the shell has been lost.

(*b*) *Lamellibranchia* (the bivalves) have two hinged shells, large gills for filter feeding and a reduced head and foot, e.g. freshwater mussel, Anadon and marine mussel, Mytilus.

(*c*) *Cephalopoda* are active pelagic animals with a reduced shell. They have a well-developed brain and include the octopus and squid.

The evolutionary affinity of the Mollusca is with the Annelida and the larval molluscs are trochophores. A link form of Mollusca, Neopilina, has many features in common with annelid morphology.

8. Echinodermata. As with the above, very little needs to be known about this phylum. Its members can be readily recognised as they are radially symmetrical and have a calcareous exoskeleton. All Echinodermata are marine and they have a water vascular system which supplies the tube-feet and is in contact with the environment.

Certain features of their development link the Echinodermata with the Chordata rather than with the Annelida-Arthropoda-Mollusca group of the invertebrates.

Divisions of the phylum are as follows:

(*a*) *Crinoidea.* The sea lilies are primitive and sessile, and are stalked when young; they are filter feeders.

(*b*) *Asteroidea.* The star fishes are five armed and are active carnivores, feeding on bivalves (*see* Fig. 12).

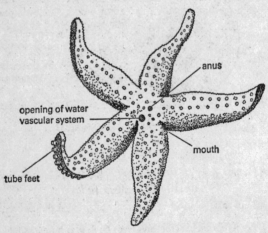

FIG. 12 *The starfish is a good example of an asteroid echinoderm.* Sea urchins, brittle stars and sea cucumbers are also members of this phylum.

(*c*) *Echinoidea.* The sea urchins are herbivores, and are adapted for rasping off encrusted algae from rocks. They are slow moving and tend to be spherical.

VERTEBRATES AND THEIR ALLIES

9. Chordata. These include invetebrate sub-phyla such as the Hemichordata (acorn worms), the Urochordata (sea squirts) and the Cephalochordata (Amphioxus). The term also includes the craniate or vertebrate animals, a single sub-phylum divided up into classes including the Pisces, Amphibia, Reptilia and Mammalia (this dominant sub-phylum is considered more fully in 10 below).

All chordates possess, at some time in their life history, a number of gill slits in the pharynx, a dorsal skeletal rod, the notochord and a tubular dorsal central nervous system.

The invertebrate Chordata are not of much importance but they do provide material which can be used to illustrate certain biological points and principles. These are as follows:

(a) The close similarity between Amphioxus and the larval lamprey indicates how fishes may be related to other Chordata.

(b) The mechanisms of filter feeding used by the sea squirts and Amphioxus illustrate some of the principles of this particular means of catching food (see VII, 15).

(c) The lower chordates show an embryological, and, in the case of the acorn worm, a larval development which relates the Chordata with the Echinodermata rather than with the Annelida-Arthropoda-Mollusca invertebrates.

(d) The larval sea squirt is somewhat similar to the adult Amphioxus and provides a possible clue as to how the free-swimming active "fish-like" animals evolved from the sessile and radially symmetrical sea squirts.

10. Craniata or Vertebrata. This is a very large and dominant sub-phylum. Besides the chordate characteristics described above, its members also have a brain with a surrounding skull, the notochord being replaced by a bony or cartilaginous set of vertebrae, a ventral heart and a post-anal tail.

(a) *Agnatha* are the most primitive Vertebrata and have no jaws. They comprise a very ancient fossil group that flourished in the Silurian period and contain the modern lampreys. The latter are of interest because they are "living fossils", as well as being one of the very few parasitic vertebrate animals.

(b) *Pisces* (fishes) are aquatic vertebrates with fins and a tail and are modified for swimming. They tend to be streamlined in shape, they have gills and a body covered with scales. A lateral line is present for detecting low-frequency vibrations.

NOTE: Fishes are further divided according to the nature of the skeletal material. Chondrichthyes have a skull and backbone of cartilage and the dogfish is a good example. Osteichthyes are bony fish, as for example the herring. Of the bony fish, the Teleostei, with cycloid scales and an air bladder (or swim bladder), are by far the most numerous of all modern fishes. From the point of view of vertebrate evolution, however, another smaller group which comprises the lung fishes and coelocanths is of interest. This latter group flourished in lakes of the Devonian period and it is thought that they evolved lungs and limbs in response to the "needs" of their environments thus giving rise to the Amphibia.

(c) *Amphibia* have legs rather than fins (they were the first tetrapods). They have to return to water to breed and have an aquatic larva which undergoes a metamorphosis before becoming an adult. Modern Amphibia have a permeable skin. The Amphibia flourished in the Carboniferous period and in the succeeding Permian period gave rise to the reptiles which are better adapted for land life.

Modern Amphibia are a reduced class and consist of frogs, toads, newts and salamanders. The terrestrial stages have lungs instead of gills and a partly divided heart.

(d) *Reptilia* have impermeable skins and a shell around the egg. The embryo forms a special amnion within the egg and reproduction takes place on land.

The reptiles were dominant vertebrates during the Mesozoic period, particularly the various groups of Dinosauria which became adapted to all sorts of environments. One branch of the dinosaurs gave rise to the Aves (birds) while another line leading from the stem Reptilia produced the Mammalia.

Modern reptiles are represented by snakes, lizards, crocodiles, turtles and tortoises. The diversity and number of species is very much reduced in comparison with the past, as the group has been largely replaced by their descendants, birds and mammals. The reptiles are of interest in showing many intermediate stages in structure between the amphibians and the higher vertebrates.

(e) *Aves* (birds) resemble reptiles in that they have a shelled egg and possess scales, etc., but are characterised by feathers and wings. Like mammals they are warm blooded. Birds are bipedal, the centre of gravity lying over the hind legs. Both auricles and ventricles are completely divided and the aortic arch is present only on the right side. The brain is large although the mass of the forebrain is composed of the ventral corpora striata rather than the dorsal cerebral cortex as seen in advanced mammals.

Associated with their large brains birds show very elaborate behaviour patterns especially in connection with their reproduction. Nesting and feeding of the young are aspects of the parental care which is very highly developed.

(f) *Mammalia* are dominant animals on land and are also successful in the sea, e.g. whales, seals and dolphins, and in the air, e.g. bats. They are warm blooded and covered with hair which provides insulation. Associated with efficient gaseous exchange a diaphragm is present. The heart is divided and a left systemic arch is present. In the skull of mammals there are three ear ossicles called the malleus, incus and stapes. Different types of teeth are found.

The reproductive processes of mammals are very advanced and the young is retained inside the female uterus and nourished via the placenta. As in reptiles and birds an amnion is present around the embryo. Parental care is well developed and after birth the young are fed with milk from the mammary glands of the mother.

The lower mammals include the egg-laying monotremes such as the duck-billed platypus—which again are of interest as "living fossils"—as well as the pouched marsupials. In the eutherian or placental mammals the following orders are included:

(*i*) Insecivora: shrews, hedgehogs.
(*ii*) Rodentia: rats, mice.
(*iii*) Lagomorpha: rabbits.
(*iv*) Carnivora: cats, dogs, bears, seals.
(*v*) Artiodactyla: cows, sheep, pigs.
(*vi*) Perissodactyla: horses.
(*vii*) Cetacea: whales, dolphins.
(*viii*) Proboscidia: elephants.
(*ix*) Chiroptera: bats.
(*x*) Primates: monkeys, man.

The cerebral cortex is large and much folded especially in the higher mammals which have the ability to learn and, in some cases, to display intelligence. The latter (*see* XI, **34**) is particularly highly developed in the human species.

PROGRESS TEST 2

1. Which sub-divisions are there in the phylum Protozoa? **(1)**

2. To which sub-phyla do Amoeba, Paramecium and Monocystis belong? **(1)**

3. By which characteristics could you recognise a coelenterate? **(2)**

4. Which important groups of parasites are found in the phylum Platyhelmintha? **(3)**

5. How could you distinguish an annelid from a nematode? **(4, 5)**

6. Which are the sub-divisions of the Arthropoda? **(6)**

7. Give two examples of molluscs. **(7)**

8. Which features are common to the Echinodermata? **(8)**

9. Distinguish between the Vertebrata and the Chordata. **(9, 10)**

10. What are the major differences between amphibians and reptiles? **(10)**

11. List eight characteristics of mammals. **(10)**

Examination questions on classification are to be found at the end of Chapter III.

Plant Classification

METHODS OF CLASSIFICATION

1. A note on plant classification. The classification of plants is generally less "tidy" than that of the animals and there is considerable doubt about the evolutionary validity of the major divisions. Thus the Algae, whose individual members show the widest diversity, may well have several evolutionary origins. It is not even clear whether mosses and liverworts (bryophytes) gave rise to the ferns and their allies (pteridophytes), or were descended from them. The relationships between the families of higher plants is also very difficult, if not impossible, to interpret.

Certain rules of nomenclature exist such as the addition of the suffixes *-phyta* to divisions, *-phyceae* to classes, *-ales* to orders, *-aceae* to families etc., but even these are not adhered to where long usage has made familiar another term, e.g. Algae and Fungi are used for classes.

NOTE: The classification of plants used here follows the scheme in current use in the Cambridge University Botany Department. It has recently been revised and is very suitable for Advanced Level requirements.

2. An outline classification of plants.

  (*a*) *Thallophyta:*

    (*i*) Bacteria

    (*ii*) Algae

        (1) Chlorophyceae (green algae)
        (2) Bacillariophyceae (diatoms)
        (3) Dinophyceae (dino-flagellates)
        (4) Phaeophyceae (brown algae)
        (5) Rhodophyceae (red algae)

    (*ii*) Algae

        (6) Euglenophyceae
        (7) Cyanophyceae (blue-green algae)

(*iii*) Fungi	(1) Phycomycetes
	(2) Ascomycetes
	(3) Basidiomycetes

(*b*) *Cormophyta:*	
(*i*) Bryophyta	(1) Musci (mosses)
	(2) Hepaticae (liverworts)
(*ii*) Pteridiophyta	(1) Filicales (ferns)
	(2) Lycopodiales
(*iii*) Spermatophyta	(1) Gymnospermae (conifers, etc.)
	(2) Angiospermae (flowering plants)

BACTERIA, ALGAE AND FUNGI

3. Bacteria. (*See* Fig. 13.)

(*a*) *Characteristics.* The smallest organisms having cellular organisation, ranging from $0\cdot5\mu$m to 8μm. DNA is present but this

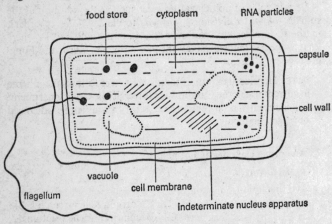

FIG. 13   *A generalised diagram of a bacterium.*

is not organised into a nucleus as in other living things. There are no large vacuoles. Spherical bacteria are called *cocci*, rod-shaped ones *bacilli* and spiral ones *spirochaetes*.

Many bacteria form resistant spores in unfavourable conditions and in this way they are dispersed. The normal means of reproduction is by fission and as this may occur every twenty minutes large populations can rapidly be built up. In some species genetic exchange of material takes place by a form of conjugation.

(*b*) *Importance*. The majority of bacteria are saprophytes and between them they possess a very wide range of enzymes which are able to digest all types of organic material.

In the soil and on the sea-bed bacteria and other organisms bring about the process of decay and thus return the bound nutrients of protoplasm back to the external medium. Certain bacteria such as Rhizobium live symbiotically with legumes and are able to fix atmospheric nitrogen. Freeliving Clostridium and Azotobacter also fix nitrogen. Nitrosomonas and Nitrobacter turn ammonia to nitrite and nitrite to nitrate respectively and are important in the nitrogen cycle.

Some bacteria are pathogenic and cause diseases in other organisms. Such bacteria may excrete toxins which are extremely poisonous to their hosts. One of the most deadly poisons known is botulin toxin of which as little as 0·0001 g may kill a man.

4. Algae: introduction. The Algae are a very diverse and extensive group of plants which show a great range of somatic organisation and complexity of reproduction and life history. It is thought that all algae must have descended from the first unicellular and photosynthetic organisms of the early seas. Within the group are to be found trends from unicellular to multicellular and from isogamy to oogamy as well as all types of life cycle from the haploid dominant (e.g. Spirogyra) to the diploid dominant (e.g. Fucus).

Algae tend to have the following characteristics:

(*a*) They are phosynthetic or holophytic organisms although a number of photosynthetic pigments may be present with the chlorophyll. The chloroplasts tend to be very large.

(*b*) *Cellulose cell walls* and *vacuoles* are present as in higher plants.

(*c*) The reproduction usually involves a *free-swimming gamete* or spore stage and for this and other reasons the Algae tend to be *confined to fresh water* or *the sea*. (There are a good many exceptions to this as for example Pleurococcus which grows on tree trunks.)

(*d*) The Algae may be *single-celled* forms or may have a *filamentous* organisation. The most advanced algae have a parenchymatous thallus with considerable differentiation of cell types.

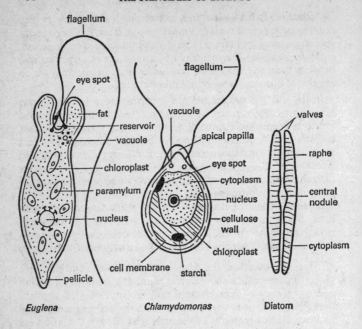

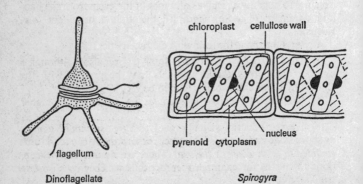

FIG. 14 *A variety of Algae.*
Note that *Euglena* appears here as a unicellular plant.

5. Algae: internal classification.

(a) *Chlorophyceae* (green algae). These are the most primitive class of the Algae and their lowest members can scarcely be distinguished from the flagellate protozoans. They are mainly freshwater forms. Pyrenoids are present in the chloroplast and the storage substances are starch or fat. Included in the Chlorophyceae are the Conjugales of which Spirogyra is an example (*see* Fig. 14). The Conjugales possibly have an affinity with certain simple fungi which also reproduce sexually by a process of conjugation. The filamentous green seaweed Enteromorpha and the flattened Ulva or "sea-lettuce" which are both found in the upper-shore zone are members of the Chlorophyceae.

An interesting freshwater order is the Volvocales which show an almost complete series of somatic and reproductive habits from the simplest to most complex members. It is thought that there are preserved in this particular order some of the intermediate stages in the evolution of the multi-cellular from the unicellular, oogamy from isogamy and the soma (or body) from the germ (or reproductive) cells. These trends are illustrated in Table I.

TABLE I.   EXAMPLES OF SPECIES IN THE ORDER VOLVOCALES

Species	Number of cells	Type of reproduction
Chlamydomonas (most species)	1	Isogamous
Chlamydomonas braunii	1	Heterogamous
Gonium	c. 16	Isogamous
Pandorina	c. 16	Isogamous to heterogamous
Eudorina	c. 32	Oogamous
Volvox	c. 20,000	Oogamous

(b) *Bacillariophyceae* (diatoms). These consist of single cells or chains; each cell is composed of two valves one of which fits inside the other in the manner of a pill box. They have chlorophyll and a brown pigment, fucoxanthin, which gives them their colour. There are a great number of species of diatoms and they make up the major part of the phytoplankton of the sea and freshwater lakes. An example of a diatom is Coscinodiscus which is found in the North Sea and provides the first food of the young plaice larva (*see* Fig. 14).

(*c*) *Dinophyceae* (dinoflagellates). These occupy a similar ecological position to diatoms and are important members of marine and freshwater plankton. Some dinoflagellates are holozoic or parasitic (*see* Fig. 14).

(*d*) *Phaeophyceae* (brown algae). These, together with certain green and red algae, make up the many species of the "seaweeds". Besides chlorophyll the brown pigment fucoxanthin is present; polysaccarhides other than starch are stored and the thallus tends to be large and well differentiated. Typical seaweeds such as the Fucus species have a holdfast which anchors them to the substratum, a long stipe and flattened lamina which bears the reproductive conceptacles. Within the lamina are photosynthetic cells as well as those specialised for storage, support and conduction (*see* Fig. 15).

In some brown algae such as Laminaria, alternation between

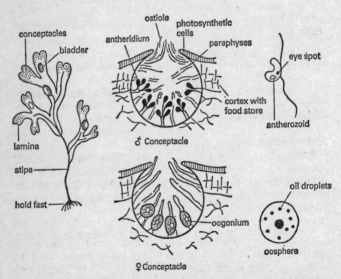

FIG. 15 Fucus vesciculosus, *the bladder wrack, showing details of the male and female conceptacles.*

The degree of differentiation is much higher than in the green algae.

haploid and diploid generations is found but in the Fucus species the diploid generation is completely dominant and the haploid exists only as a vestigial and transitory phase. In this respect these algae resemble the higher plants in their life cycles and it may be presumed that the same advantages that have led to selection of the diploid state have operated in the evolution of both groups.

The brown algae show good examples of the strict zonation that exists on the sea shore. Each species appears adapted to a specific zone according to the extent of desiccation etc., that it can withstand. The male sex cells (antherozoids) and female sex cells (oospheres) are liberated from the ripe conceptacles and the oospore that results from fertilisation drifts in the sea. If it happens to germinate in the correct zone the young plant survives.

(e) *Rhodophyceae* (red algae). This is another group of sea-weeds which appear red owing to the pigment phycoerythrin. They have a large and complex thallus and elaborate life histories. Unlike the brown algae the dominant phase of the plant is haploid.

(f) *Euglenophyceae*. A small freshwater group which includes Euglena, its systematic position is somewhat difficult to determine as its members have various animal-like features and are included by zoologists as flagellate protozoans. Some colourless forms may feed heterotrophically.

(g) *Cyanophyceae* (blue-green algae). These are another rather anomalous group possibly related to the bacteria. Single and multicellular forms exist but the organisation is very simple and no chloroplasts or vacuoles are present. Phycocyanin gives them their colour. They have no definite nucleus.

Blue-green algae are found in the soil and freshwater habitats and one of the species that grows in water is Nostoc which is able to fix nitrogen. Nostoc grows in the rice fields of China and its ability to fix atmospheric nitrogen allows the fields to retain their fertility year after year.

6. Importance of algae. The main important functions of Algae are as follows:

(a) *As primary producers.* Algae occupy the same place in freshwater and marine food chains as does grass on land. All food chains in water tend to be derived from the algae of the plankton and the usual arrangement can be shown as in Fig. 16.

(b) *As oil producers.* Algae play a major role although the

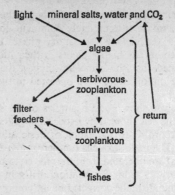

FIG. 16 *A marine food chain.*

process takes many millions of years to achieve. In certain areas of the sea, vast overproduction of phytoplankton is followed by its death and sinking to the sea bed. Under the correct conditions these masses of plants give rise to oil and gas deposits.

(*c*) *Evolutionary significance.* Study of the Algae is important in throwing light on the simplest types of plants and the relationships of these to animals and fungi. The evolution of complex organisation in the soma, of elaborate life cycles and sexual processes is all clearly illustrated within the algae.

7. Fungi: introduction. The Fungi are a widely successful group of plants, thought to have been variously derived from the Algae. They tend to have the following characteristics:

(*a*) *No chlorophyll is present* and Fungi are either *saprophytic*, feeding by rhizoids, or else *parasitic*, feeding by haustoria.

(*b*) The body is made up of fine threads called hyphae which may or may not have cross walls. A large number of threads make up the *mycelium* which in turn may be diffuse, as in Mucor, or massive, as in mushrooms.

(*c*) The food store is composed of *droplets of oil* and the polysaccharide *glycogen*.

(*d*) Cellulose is replaced by *fungal cellulose* which has similar structural properties but is a different molecule from normal plant cellulose.

(*e*) A resistant phase, the *spore*, is present in the life cycles of Fungi and the different types of spores give the basis of classi-

fication within the group. Spores are often held above the substratum by a special *sporangiophore* and this helps in their
dispersal.

(*f*) Fungi are *permeable to water* and normally live in damp
places or in water. They are very susceptible to desiccation.

8. Fungi: internal classification.

(*a*) *Phycomycetes* are the most primitive group of the Fungi.
They have a coenocytic or non-septate mycelium and reproduce
asexually by the cutting off of spores from the end of hyphae
(*see* Fig. 17). Where these spores are equipped with flagella and
can swim they are termed zoospores but where they are resistant,
aerially dispersed forms they are called conidiospores. Sexual reproduction takes place in one of two ways and the Phycomycetes
are sub-divided on this basis:

(*i*) *Oomycetes* have antheridia and archegonia. The small
antheridium grows near to the larger archegonium and a fertilisation tube forms between. A single male nucleus passes into
the oosphere which becomes an encysted zoospore. The oomycetes include the following series of saprophytes and parasites:

TABLE II. EXAMPLES OF SPECIES IN THE SUBCLASS OOMYCETES

Disease	Species	Host(s)	Asexual reproduction
"Fish spot"	Saprolegnia	Freshwater fishes	Zoospores
"Damping-off"	Pythium	Seedlings	Zoospores or Conidia
Blight	Phytophthora	Potatoes, etc.	Zoospores or Conidia
"Downy mildew"	Peronspora	Cruciferae etc.	Conidia
"Downy mildew"	Cystopus	Cruciferae, etc.	Conidia

The series shows increasing adaptation to survival on land
hosts and some of its members are of great economic importance,
such as the Phytophthora infestans (the potato blight).

(*ii*) *Zygomycetes* reproduce sexually by conjugation between
two similar hyphae. The process is somewhat similar to that seen

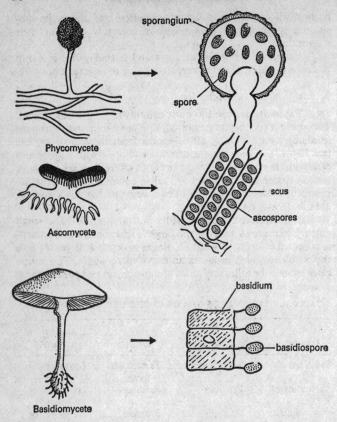

FIG. 17 *The Fungi are sub-divided into groups according to the type of spores that they produce.*

The Phycomycetes are probably the most primitive group.

in Spirogyra but the two filaments put out pro-gametangia which meet in the middle and form a thick-walled resistant zygospore.

Moulds such as Mucor and Rhizopus belong to the Zygomycetes.

(*b*) *Ascomycetes* have hyphae that are divided up by cross-walls. The ascospore characteristic of the group is produced by a fusion of nuclei and subsequent reduction division within the ascus, a flask-shaped reproductive body (*see* Fig. 16).

Many asci are present on a single fruiting body and it is the shape of the latter that allows further sub-divisions. Among the important Ascomycetes are Erysiphe and Claviceps, which are parasites of cereals, and the beneficial fungi Saccharomyces (yeast) and Penicillium. Yeast is described more fully in VII, **19**. It should be noted that it does not easily reproduce sexually and in many ways it is an aberrant and atypical ascomycete.

(*c*) *Basidiomycetes* have hyphae with cross walls and the collection of hyphae, or the mycelium, may be built up into a massive body with some degree of differentiation as in mushrooms and toadstools (*see* Fig. 17).

The basidiospores, from which the group derives its name, are formed by nuclear fusion and subsequent reduction division in the basidium. Four basidiospores are thus formed and each of these is nipped off from the basidium, being retained by a thin sterigma. Thes napping of this is due to a surface-tension mechanism which leads to the spore being released.

Certain basidiomycetes, such as rusts and smuts, are very harmful parasites of cereals while others such as Stereum parasitise trees. Mushrooms, e.g. Psalliota, and toadstools are saprophytic and play a role in the breakdown and circulation of organic matter.

9. The importance of fungi.

(*a*) *As saprophytes*, many fungi act with other micro-organisms in the breakdown of organic matter in the soil and the return of nutrients into circulation. There may be as much as 2 m of fungal hyphae in 1 cm^3 of soil.

(*b*) *As parasites of plants* fungi are of great economic importance. Blights, mildews, smuts and rusts and other infections cause damage or total loss to many of the food plants grown by man. There are an ever-increasing number of fungicidal sprays and application of these at the correct times together with steps to reduce the chances of infections does much to reduce losses.

(*c*) *As parasites of animals* fungi are of less importance than bacteria and viruses. In fresh water, fungal diseases such as Saprolegnia attack a variety of fishes. Man suffers from certain fungal infections of the skin such as ringworm, and tinea infections such as athlete's foot. There are a few fungal infections of the lungs, genital passages and deeper parts of the body.

(*d*) *Industrially* fungi, usually yeast, are important in brewing and breadmaking. The biochemistry of fermentation is described in VIII, 6.

(*e*) *Antibiotics* such as penicillin, tetramycin, ledermycin, etc., are derived from the secretions of various soil fungi. These secretions are toxic to a large number of types of bacteria and their use, as antibiotics, has done much to revolutionise our treatment of diseases due to bacteria in recent years.

BRYOPHYTA, PTERIDOPHYTA AND SPERMATOPHYTA

10. Bryophyta. This is the division of the plant kingdom which includes the mosses and liverworts (*see* Fig. 18). Its exact relation-

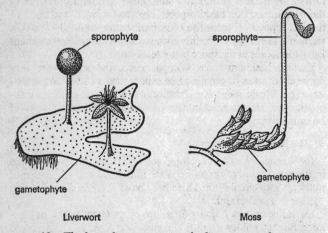

FIG. 18 *The bryophytes comprise the liverworts and mosses.* The whole group is still very dependent on water for reproduction.

ships with the divisions on either side of it are not very clear and there are some reasons for thinking that it was derived from Algae and gave rise to pteridophytes and others, which seems to indicate that it is a somewhat degenerate descendant of the pteridophytes.

Bryophytes show alternation of generations and tend to have a dominant gametophyte and a parasitic or semiparasitic sporo-

phyte that grows from it. The group usually needs water for fertilisation to take place and having no cuticle the thallus easily
dries up. True roots are lacking and chains of single-celled
rhizoids are not able to penetrate far into the soil so that, for all
the above reasons, bryophytes usually grow in wet places. The
chloroplasts are large and disc shaped. Asexual reproduction is
common and takes place by production of gemmae.

Sub-divisions of the Bryophyta are as below:

(a) *Musci* (mosses). Gametophyte differentiated and "leafy".
The sporophyte is persistent with a complex capsule and with
peristomal teeth for spore dispersal. The sporophyte is partly
photosynthetic. Examples are Funaria, Fontinalis and Poly-
trichum which are common species in most countries and Sphag-
num, the larger "bog-moss", which grows in great quantities as a
dominant plant in some parts of Ireland and elsewhere.

(b) *Hepaticae* (liverworts). The gametophyte is thalloid and
much more primitive in appearance than that of the moss. The
rhizoids are made up of only one cell. The sporophyte lasts only a
short time and is not normally able to photosynthesise. It dis-
perses its spores by elators, which are cells within the capsule
which twist as they dry out.

Liverworts are more dependent on water than mosses and may
be found along the banks of streams and in shady, damp places.
A common species is Pellia.

The bryophytes are mainly of interest because of their life
cycles and a description of these together with an indication of
their relationships to the life cycles of other plants is outlined in
X, 6.

11. Pteridophyta. These include the ferns, horse tails and lyco-
pods. Like the Bryophta, they have alternation of generations
but in their case the sporophyte is dominant and the gametophyte thalloid is very much reduced. Differentiation is well marked in the sporophyte whose leaf structure, at least internally, is
like that of the flowering plants. There is a stele with xylem and
phloem and a cuticle over the surface of the leaves and stem.

Fertilisation is by the swimming of motile antherozoids from
the antheridium to the archegonium and oosphere (*see* Fig. 19),
so that pteridophytes need external water for reproduction. On
the whole, however, they are far less dependent on water than
bryophytes.

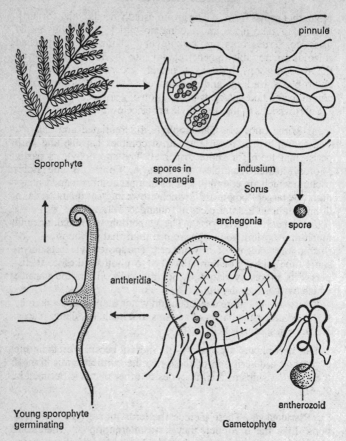

FIG. 19   *The life cycle of a pteridophyte fern.*

Among present-day survivors of this once large and dominant group (coal is made mainly from fossilised pteridophytes), are the ferns such as Dryopteris. A few species of these, namely the water ferns, are heterosporous, as is the lycopod Selaginella.

Heterospory is of interest in indicating the way in which the life cycles of the higher plants may have evolved from the pteridophytes. This problem is discussed in X, **6, 7.**

12. Spermatophyta. These are the seed plants that make up the present-day dominant flora of the earth. The seed itself is an integumented megasporangium with a single megaspore (*the ovule*) and within this a vestigial female-gametophyte generation develops. The seed is retained and nourished by the sporophyte during its development.

The microspore or pollen grain is transferred to the proximity of the ovule by wind or insects, a process called pollination and one much more suited to land life than having a swimming gamete. On germination the pollen grain puts out a pollen tube which may be considered as gametophyte tissue. Two male nuclei or motile antherozoids in some gymnosperms are formed and transferred through the micropyle into the ovule. In angiosperms one of these fuses with the female nucleus and the other forms a triploid endosperm nucleus with the polar or secondary nuclei of the ovule. Storage tissue, either in the endosperm or within the first leaves, the cotyledons, is present in the seed which becomes dispersed by one means or another from the parent plant.

The organisation of the spermatophyte soma is also more elaborate than that of pteridophytes and the stele may contain xylem vessels which allow rapid conduction of water. Spermatophytes are very well adapted for land life and have even been able to colonise such unlikely habitats as deserts and sand-dunes.

The seed plants are sub-divided as follows:

(*a*) *Gymnospermae* have "naked seeds", no carpel being present. No vessels are present in the xylem tissue. Various fossil groups are known with affinities to the pteridophytes but the major surviving group is the Coniferae. These are characterised by their needle-like leaves and cones. Pines and fir trees are examples of conifers.

(*b*) *Angiospermae* have seeds which are enclosed in the carpel which normally become modified to assist dispersion. The double fertilisation process outlined above takes place and an endosperm tissue forms within the ovule. Vessels are present. A scheme of classification of angiosperms is seen in Fig. 20.

VIRUSES

Viruses are included here for convenience. It does not imply that these organisms are specifically related to plants.

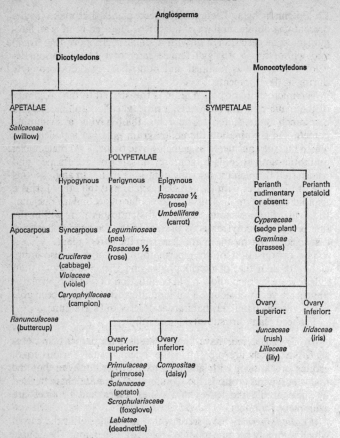

FIG. 20 *A scheme of classification of angiosperms according to floral anatomy.*

The key of the ordinary flora is based on a scheme such as the one shown above. Note that the Rosaceae are a divided family; one half is perigynous, the other half epigynous.

13. Size. Viruses are exceedingly small ranging from 20 to 400 nm. Because of their small size they are able to pass through the porcelain filters that restrain bacteria. It is also impossible to see them with a conventional microscope as they are below its powers of resolution. In recent years it has been possible to see and photograph fixed preparations of viruses, using the electron microscope.

NOTE:

$$1 \; \mu m = 10^{-6} m$$
$$1 \; nm = 10^{-\mu} m$$

14. Structure. Many viruses appear to consist of an outer protein envelope which encloses nucleic acid. The virus is thus mainly composed of genetic material and has none of the other structures associated with cells. Viruses called *bacteriophages* may have a tail-like structure, which is important in gaining entry to a host bacterial cell. Some viruses can be made to form crystals, showing that they are pure protein molecules.

NOTE: These "T" phages have been of much use in the elucidation of the genetic code and will be discussed in this context in V, 21.

15. Reproduction. In bacteriophages the virus nucleic acid enters the host cell via specific receptor sites on the cell wall, leaving the empty protein case on the exterior. After a latent period of some ten minutes the bacteriophage appears to take over the protein-synthesis mechanism of the host causing its microsomes to produce 'phage nucleic acid and new protein envelopes. The energy required for this is derived from the mitochondrial activity of the host itself.

Within hours (or minutes, according to the virus–host system concerned), a number from twenty to two hundred new and complete 'phages are manufactured by the host cell which then undergoes *lysis* or breakdown. The 'phages are thus released and may infect new cells.

16. Importance of viruses. All viruses are parasitic and they are collectively the cause of very many animal and plant diseases. In man, examples of virus infections include warts, measles, small-pox and chicken-pox, poliomyelitis, rabies, influenza, a type of pneumonia, yellow fever and dengue fever and some forms of enteritis. Foot-and-mouth disease and myxomatosis are also virus infections. In plants many types of mosaic diseases (e.g. in tobacco and potatoes) are due to viruses. It also appears that certain comparatively rare forms of cancer found in parts of Africa and elsewhere are due to viruses.

17. Resistance to viruses. As viruses do not have the parts of the cell and the mechanisms against which antibiotics 'operate, they are (except for some of the very large ones) resistant to these drugs. The living cell produces an antiviral agent called *interferon* and this substance seems to have a common structure wherever it is found. It has only recently become possible to use commercially manufactured interferon against virus infections.

Outside the bodies of their hosts viruses are easily destroyed by heat and by disinfectants but they may be resistant to cold. Viruses do not form spores and their life outside the bodies of the host tends to be very short. They are very susceptible to ultraviolet light.

18. Other phenomena associated with viruses.

(*a*) Certain viruses show the phenomenon of *reduction* whereby the virus particles become permanently incorporated with the genetic mechanism of the host and produce inherited abnormalities in the daughter cells.

(*b*) *Transduction* is another phenomenon associated with certain viruses by which they are able to transfer into the cells they infect genetic material from previously infected cells. This property was used in experiments to demonstrate that nucleic acids did, in fact, carry the heredity code.

(*c*) *Prophages* are viruses that infect host cells but remain dormant in the latter for many years, although under some conditions, as yet not very well known, they may attack and destroy the cells giving rise to the symptoms of the disease.

PROGRESS TEST 3

1. What are bacteria? (3)
2. How are the Algae sub-divided? (5)
3. What does "zonation" mean when applied to seashore algae? (5)
4. How may the Fungi be classified? (8)
5. What is the difference between a zoospore and an oospore? (8)
6. Name two ascomycete fungi of economic importance. (8)
7. What is the importance of the Fungi? (9)
8. How are the mosses distinguished from the liverworts? (10)

9. What does "alternation of generations" mean when applied to pteridophytes? **(11)**

10. What does the word gymnosperm mean? **(12)**

EXAMINATION QUESTIONS

These questions cover the material found in Chapters I to III.

The figures in **bold** type indicate the marks allocated to each question or part question.

1. Give an account of the characteristic features of a virus and explain the difficulty of classifying viruses as living organisms. **(9)**

Point out the main ways in which the structure of a bacterium differs from the structure of a virus. **(9)**

(Total **18**) *Cambridge,* 1977

2. Complete the table below to show details of characteristics of classes in the phylum *Arthropoda.*

Class		Crustacea		
Number of main divisions of the body			two	
Organs of gaseous exchange	tracheal tubes			
Number of pairs of legs				one or two per segment

(Total **9**) *part question, Associated Examining Board,* 1976

3. Some animals are triploblastic and have a gut with a single opening.

(*a*) *Name* the phylum to which these animals belong............

(*b*) *Name one* of the classes in this phylum.........................

(*c*) Give *one* characteristic which distinguishes the members of this class from the others in the phylum.

(Total **3**) *part question, Associated Examining Board,* 1976

4. Sort out the following organisms into 4 groups according to their evolutionary affinities: *Chlamydomonas,* ragworm, tape-

worm, *Fucus*, leech, moss, *Planaria*, *Spirogyra*, liverfluke, liver-wort, earthworm.

Group

Group

Group

Group

Select one of the animal groups and complete the following:

Group selected............... (Indicate number of the group.)

Name the phylum to which all organisms in the group belong.

....

State *three* important characteristics of the phylum.

(*Total* 8) *Associated Examining Board*, 1977

5. Why are fishes, birds and mammals classified in the same phylum?

What features distinguish mammals from the other verte-brates?   **(10, 10)**

London, 1976

6. Review the range of form in any *one* animal phylum.   **(25)**

London, S level, 1976

7. Write a general account of *one* of the following: (*a*) bacteria and viruses; (*b*) insects.

after London, 1975

8. Give an account of the methods used to classify plants and animals.

after London, 1975

CHAPTER IV

The Chemistry of Biological Compounds

BIOCHEMISTRY

1. The relationship between biochemistry and general biology. In I, 1, it was explained how there has been a change in the emphasis of contemporary biology. This change is reflected in current research interests and courses at universities and to a lesser extent in the content of the new Advanced Level syllabuses. Underlying this shift of emphasis has been the development of physical, chemical and even mathematical techniques whereby many biological processes can be followed more quantitatively than was possible before. *Biochemistry* means the chemistry of living things, and the study of metabolic functions such as energy exchange, protein synthesis, active transport, etc. at a molecular level is a major part of modern biological research.

In order for the present-day student to understand the comparative physiology that makes up much of the new syllabuses it is necessary for him to have an idea of the basic chemistry of biological compounds.

Ultimately it will be seen that every metabolic function is the result of changes taking place within, or at the surface of, living protoplasm. These changes are largely biochemical and the two disciplines of physiology and biochemistry are becoming less distinguishable each year.

2. Some important techniques of biochemistry. Many of the research techniques used by biochemists are far outside the scope of Advanced Level, although some, as for example manometry and chromatography are finding an increasing use in practical courses at this level. A very brief outline of some of the most widely used methods is given below:

(*a*) The *p*H of a solution may be accurately measured by a meter which separates different H^+ concentrations across a thin glass membrane. A potential difference related to $[H^+]$ is set up and, after amplification, may be read off as an actual *p*H value.

(*b*) Biochemical compounds, like other chemicals and elements, have a property of showing maximum absorption of light of different wavelengths. Very small (i.e. μg) amounts of biochemical substances can be quantitatively determined by measuring the percentage absorption they show at selected wavelengths as compared with a solution of known concentration. Substances such as adenosine triphosphate can be estimated in this way as can inorganic ions. The method is called *spectrophotometry*.

(c) It is possible to separate small quantities of organic (and inorganic) compounds according to the rate they travel along a piece of paper, a column of suitable material or indeed a liquid. This allows detection and identification of minute amounts of reaction products as the rate of travel of any given substance under controlled conditions is constant and thus unknowns can be compared with standard solutions. The technique is called *chromatography*.

(d) Perhaps the most useful modern method is that involving the use of *radioactive tracer substances*, also called radiobiology. Specific isotopes allow the labelling of compounds that can be recognised at any stage in a reaction. Isotopes for most elements of biological importance are available and radioactive carbon, C^{14}, has been incorporated into many organic molecules which may be suitable for any particular investigation. A classical application of the above technique, largely combined with chromatography, was the unravelling of the pathway of photosynthesis by Calvin in 1949.

(e) Ion exchange resins are used for the *extraction* and *purification* of *organic compounds* of small molecular groups and ionisable groups. This includes such nitrogen-containing substances as amino-acids.

(f) *Ultracentrifuges* are perhaps best considered as physical rather than chemical instruments but their high speeds (over 50,000 r.p.m.) make it possible to separate substances of different densities, for example, the mitochondria, from surrounding protoplasm. The various fractions can then be examined by other means.

(g) Changes in the production or uptake of a gas such as CO_2 or O_2 may be measured very accurately by the Warburg type of *manometer*. Because many biological activities, e.g. respiration, are accompanied by changes in gas volumes the method has been used extensively in determining the rates of such processes under varying conditions. Very small changes down to 0·015 ml can be accurately assessed (*see also* VIII, 30).

(*h*) A very complex method for the analysis of protein structure is *X*-ray crystallography, the scatter patterns of these rays as they pass through a crystal of the substance being related to the density of the atoms. This technique is also of classical interest as it was used by Watson and Crick in 1953 to determine the molecular configuration of nucleic acid, the chemical of the gene.

CARBOHYDRATES

3. Definition. These are substances normally characterised by having carbon, hydrogen and oxygen in the molecules with the ratio of the hydrogen and oxygen being the same as that in water, i.e. $C_x(H_2O)_y$. There are exceptions to this, for example pentose sugars of the nucleic acid DNA have a formula $C_5H_{10}O_4$. There are also a very small number of carbohydrates that contain nitrogen, e.g. chitin.

Carbohydrates are very important, widespread biological compounds as they both provide a *source of metabolic energy* and act *structurally* in animal and plant tissues. The basic unit of the carbohydrate is the *saccharide* or sugar molecule and the group is sub-divided according to the number of saccharides contained. Thus monosaccharides contain a single sugar, oligosaccharides a number and polysaccharides up to several thousand sugars to a single molecule.

4. Monosaccharides. These include the sugars glucose and fructose both of which are based on 6-carbon atoms, therefore termed hexoses, the 5-carbon sugars or pentoses which include ribose, and the smaller 3-carbon sugars called trioses.

Monosaccharides are the most active carbohydrates and are the main respiratory substrate of all cells. They are also the form in which sugars are transported in animals, and in plants contribute to osmotic effects important in turgor.

Because monosaccharides are the building units of other more complex carbohydrates it is necessary to understand something of their molecular configurations. In the hexoses there may be structural isomers (with the empirical formula $C_6H_{12}O_6$) but different arrangements of the constituent atoms, as in Fig. 21.

NOTE: (1) For speed of writing, the modern convention is to use *stick formulae* with certain C and H groups being omitted. Thus the formulae of glucose and fructose are written as in Fig. 22.

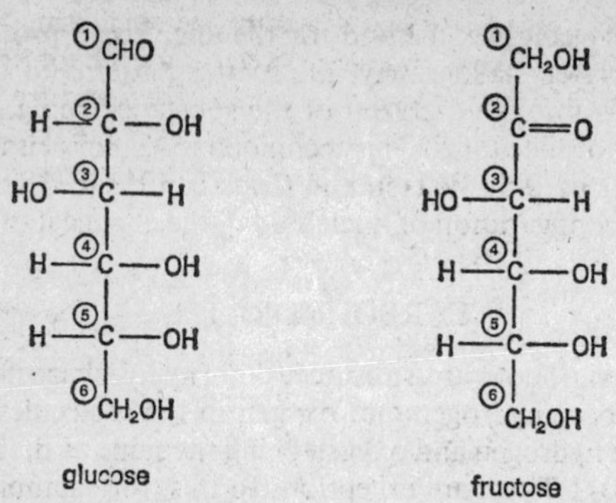

FIG. 21 *Formulae for glucose and fructose.*

glucose   fructose

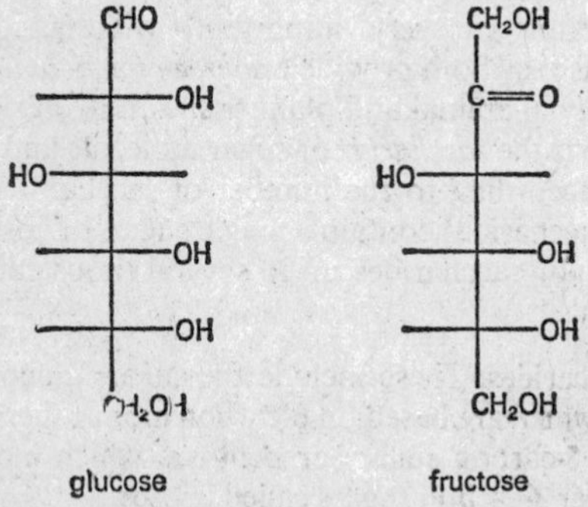

FIG. 22 *Stick formulae for glucose and fructose.*

This convention will be observed for other carbohydrates described. (2) Besides the structural isomers shown, it is possible to obtain stereoisomers of each of the above where the groups attached to the four central carbons are reversed so that mirror-image molecules result. The two stereoisomers have the property of rotating polarised light in different directions and those shown above turning it to the right are called dextrorotary or just D-molecules. The mirror-image molecules would be laevo-rotatory or L-molecules.

It is found that the two hexose sugars glucose and fructose do not normally occur in the chain form shown above but that ring structures are formed with 2-carbon atoms linking across an oxygen bridge. These ring structures may be of two forms according to which carbon atoms are involved and the forms are called *pyranose* and *furanose* respectively. Glucose usually occurs as the pyranose (*see* Fig. 23), while fructose more often occurs as the furanose (*see* Fig. 24).

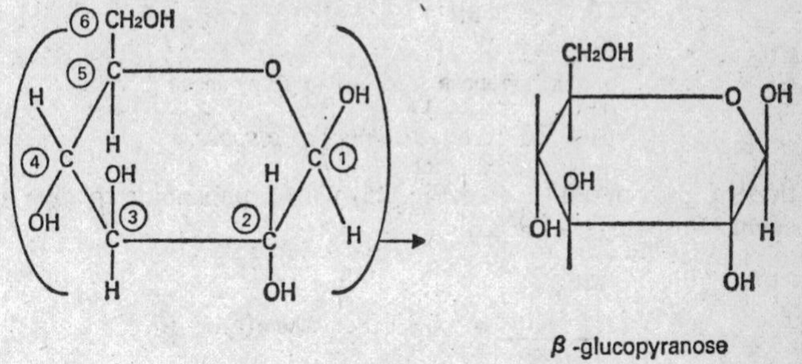

β -glucopyranose

FIG. 23   *Ring structure of glucose.*

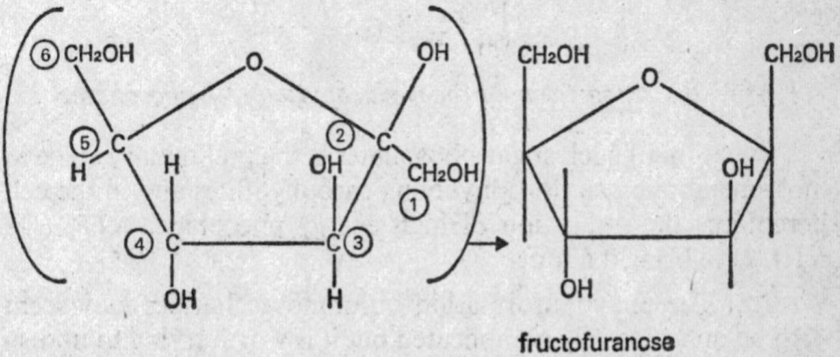

fructofuranose

FIG. 24   *Ring structure of fructose.*

Stereoisomers are also possible according to the arrangement of the groups around asymmetrical carbon atoms. Thus for glucopyranose we may have the structure indicated in Fig. 25. This point is mentioned because it is important in understanding differences between cellulose and starch.

A property of many monosaccharides is their ability to form glycosides where the —OH group of ①C reacts with other radicles. A common example of such a glycoside is the combina-

α -glucopyranose   β -glucopyranose

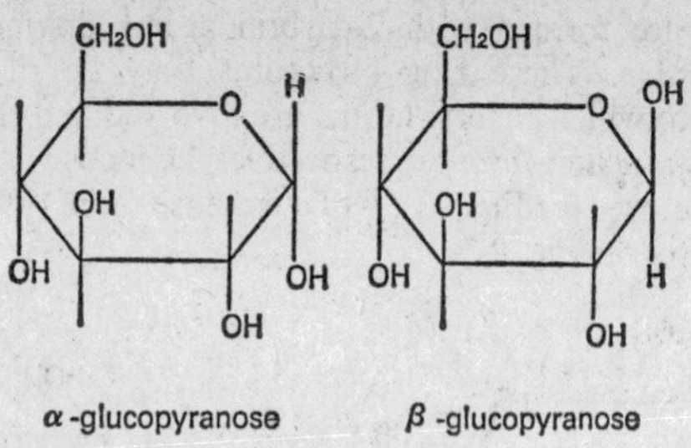

FIG. 25   *Stereoisomers of glucose.*

tion of glucopyranose (*see* Fig. 25) with a phosphate to give a sugar phospate (*see* Fig. 26).

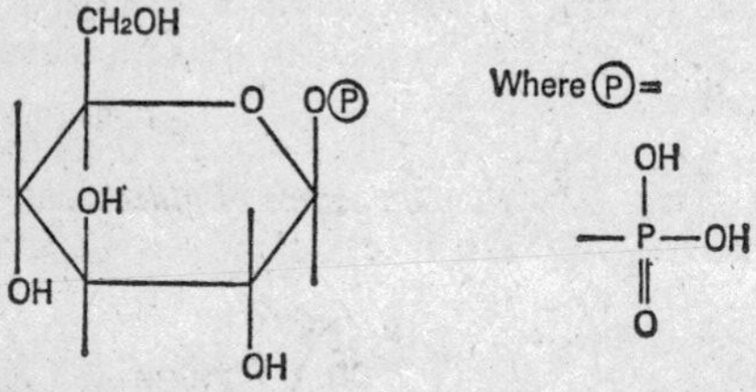

FIG. 26   *Ring formula for α-glucopyranose-1-phosphate.*

Formation of such sugar phosphates is the preliminary stage in most metabolic reactions involving carbohydrates and in the cell it requires the utilisation of high energy phosphate, ATP, (*see* VIII, 2) to bring it about.

NOTE: The above information on monosaccharides may seem to be unnecessarily complicated but it is worth trying to understand as the chemical structure and metabolism of the larger carbohydrates is based on the monosaccharide units from which they are synthesised.

5. Oligosaccharides. These include carbohydrates which on hydrolysis give a small number of saccharide units. By far the most common are the disaccharides which consist of two saccharide units and which include the sugars maltose (malt sugar), sucrose (cane sugar), and lactose (milk sugar). Disaccharides are a form of glycoside as the second sugar condenses on the —OH group of the ①C of the first.

(a) *Maltose* is formed by condensation of two glucopyranose units at the α position (*see* Fig. 27). This sugar is not of import-

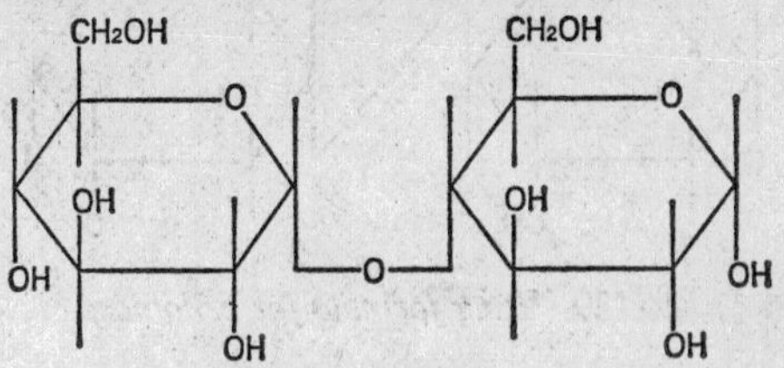

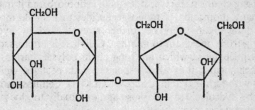

FIG. 27   *Ring formula for maltose.*

ance in itself but forms the repetitive unit in the polysaccharide starch which is very widespread as a storage carbohydrate. Maltose is hydrolysed back to two glucose molecules by the action of the enzyme *maltase*.

(b) *Sucrose* is formed by the condensation of fructofuranose and α-glucopyranose (normally as sugar phosphates). It can be represented in Fig. 28. It is a very important sugar and the main

FIG. 28   *Ring formula for sucrose.*

way in which carbohydrates are transported in plants. It is also found in plants such as sugar beet and sugar cane, as a storage product. Enzymic hydrolysis of sucrose is by *sucrase* (sometimes called invertase).

(c) *Lactose* involves the condensation of glucose and another monosaccharide called galactose. In this case a β-glycoside is formed and hydrolysis may take place by the enzyme *lactase*.

(d) *Cellobiose* is another disaccharide which, like maltose, is not usually found except as a transitory product of hydrolysis. This is the result of two glucopyranoses joining up in the β configuration (*see* Fig. 29). As this is the structural unit of the

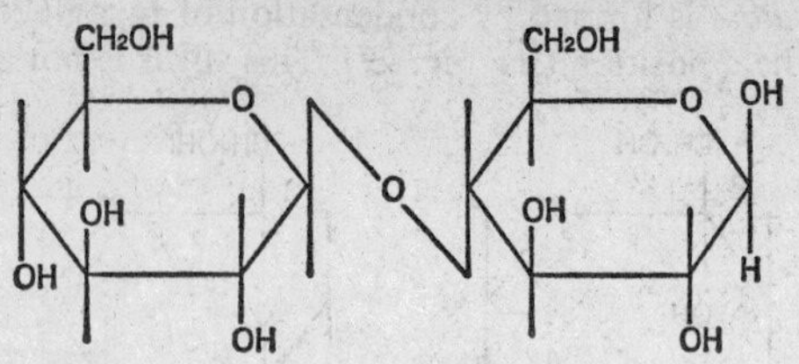

FIG. 29   *Ring formula for cellobiose.*

polysaccharide cellulose, it is also important. Enzymes, as we shall see later, are very specific in action which is the reason why those which will digest starch, e.g. our own salivary amylase, will not digest cellulose.

6. Polysaccharides. These are the largest carbohydrates and are important both structurally, e.g. cellulose, lignin and chitin, and also as storage products, e.g. starch and glycogen.

(*a*) *Cellulose*, which makes up the cell wall of plants, is made up of long chains of β-glycosides as in cellobiose and it may contain many thousands of monosaccharides to a single molecule. These long chains may be further strengthened by hydrogen bonding from one chain to another. Cellulose is often associated with hemicellulose which are long chain polysaccharides based on other sugars; it is also associated with pectates which are sugars with terminal acid groups (—COOH). Cellulose is broken down by hydrolysis to cellobiose and eventually to glucose. The enzyme *cellulase* is involved.

(*b*) *Lignin* is a structural chemical forming the wood of plants. Unlike cellulose and other carbohydrates it is made up of aromatic units which form long chains with cross-linkages. It is thought that the original units are themselves synthesised from monosaccharides. Lignin is very important in the support of woody plants.

(*c*) *Chitin*, as found in the cuticle of insects and other animals, is another structural polysaccharide. Chitin also forms the cell wall of fungi. The chitin molecule is also made up of long chains of monosaccharide units but in this case they have the amino-group, i.e. NH_2 attached. The substance can be tanned by oxygen cross-linkages forming between adjacent chains and is very tough and durable.

(*d*) *Storage polysaccharides* include *starch* and *glycogen*. The former is the result of the condensation of molecules of the α-glucopyranose-phosphate type to give chains several hundred sugars in length (*see* Fig. 30). The major constituent of starch is called *amylose* but the substance is a complex and also contains

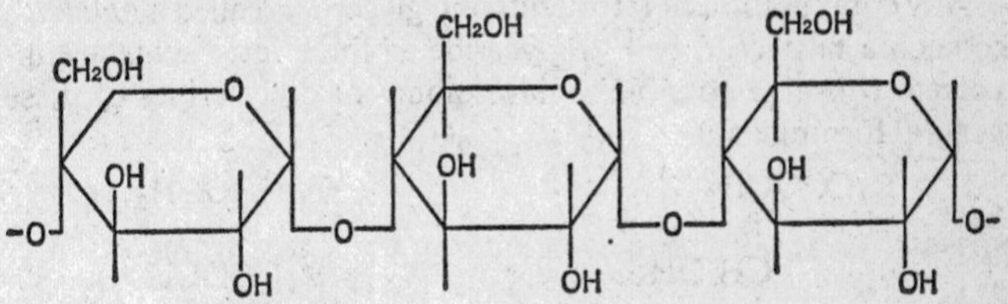

FIG. 30   *Ring formula for starch.*

amylopectin which is a branched polysaccharide with shorter chains. *Amylases* or *diastases* hydrolyse starch but these enzymes are themselves complexes, with members specific for the various linkages within the starch. The first result of hydrolysis is maltose and the final product glucose.

Glycogen is a main carbohydrate storage substance of animals and fungi and is made up of molecules rather like amylopectin but with more numerous sidechains. The structure of glycogen has been likened to a fruit tree hung with ripe fruit. Thus there is a fairly stable polymer spine (the "tree") from which jut out easily hydrolysed short chains of hexose sugars (the "fruit"). The analogy is apt and helps us to visualise the suitability of the molecule as a storage substance.

Inulin is a storage polysaccharide found in certain Compositae (the daisy family) as for example in the dandelion and dahlia. This is composed of large numbers of fructofuranose molecules condensed together in chains.

LIPIDS

7. Lipids and related compounds. The *lipids* are organic substances that contain C, H and O and where the proportion of O to the other two elements is very low. They include the fats, oils and waxes.

Fats are formed by the condensation of a number of fatty acids with the trihydroxy alcohol glycerol:

$$\begin{array}{ccc}
R_1COOH & OH.CH_2 & R_1COOCH_2 \\
| & | & | \\
R_2COOH + & OH.CH & \rightarrow \quad R_2COOCH \quad + \quad 3H_2O \\
| & | & | \\
R_3COOH & OHCH_2 & R_3COOCH_2
\end{array}$$

Any combination of fatty acids and glycerol is called a *glyceride* so such a fat would be a triglyceride as three fatty acids are involved. It is also possible to have mono- or diglycerides of these general formulae:

$$\begin{array}{cc}
R_1COOCH_2 & \qquad R_1COOCH_2 \\
| & \qquad | \\
CH.OH & \qquad R_2COOCH \\
| & \qquad | \\
CH_2OH & \qquad CH_2OH
\end{array}$$

$$\quad \textit{monoglyceride} \qquad\qquad \textit{diglyceride}$$

(*a*) *Fats or oils*. Whether or not the molecule behaves as a fat or as an oil depends on which particular fatty acids are involved. Plants tend to have liquid oils which contain unsaturated oleic acid. ($C_{17}H_{33}COOH$) whereas animals store solid fats which often contain palmitic acid ($C_{15}H_{31}COOH$) or stearic acid ($C_{17}H_{35}COOH$) which contain saturated fatty acids.

(*b*) *Waxes*. The waxes, such as are found on the leaves of plants, in the ear or as beeswax, are combinations of long chain fatty acids with monohydroxy alcohols of high molecular weight. They are very insoluble and inert substances.

(*c*) *Phospho-lipids*. Although fatty acids are normally found as glycerides, i.e. combined with glycerol, they may also condense with other chemicals. One such important combination is with a nitrogen-containing base across a phosphoric acid link. This gives a phospho-lipid, which is an essential part of the cellular and other membranes. Phospho-lipids have this general formula.

$$CH_2COOR_1$$
$$|$$
$$CH.COOR_2$$
$$|$$

$$CH_2O.P-O-CH_2\overset{+}{N}-CH_3 \quad \text{(and a balancing anion, e.g. } Cl^-\text{)}$$

with OH and CH_3 groups, P double-bonded to O, and two further CH_3 groups on N.

Lipids and related substances can be used as respiratory substrates, being passed into the Krebs Cycle a little at a time via the intermediate compound acetyl co-enzyme A.

They have a high energy yield of $38 \cdot 6$ J/g and because of this and their chemically inert nature they are useful storage materials for living organisms. Thus many insects have large fat bodies. Fat is accumulated by hibernating mammals while in plants it occurs in certain seeds such as those of sunflowers, groundnuts, castor oil plants, etc.

AMINO-ACIDS, PROTEINS AND OTHER NITROGEN-CONTAINING MOLECULES

8. Amino-acids. These contain hydrogen, oxygen, carbon and nitrogen and, in a few cases, sulphur. There are some twenty-three different amino-acids and all types of protein are made up of various combinations of amino-acids.

In solution all amino-acids behave as electrolytes, the characteristic $—NH_2$ and $—COOH$ groups becoming $—NH_3{}^+$ and $—COO'$ respectively. They are thus very reactive and can behave as acids or bases.

Chemically amino-acids may be classified according to the numbers of groups that are present or the configuration of the carbon stem.

(a) Members of this group have one $—COOH$ and one $—NH_2$:

$$
\begin{array}{ll}
\text{Glycine (Gly)} & \overset{\displaystyle NH_2}{\overset{\displaystyle |}{H—CH—COOH}}
\end{array}
$$

$$
\begin{array}{ll}
\text{Alanine (Ala)} & \overset{\displaystyle NH_2}{\overset{\displaystyle |}{CH_3—CHCOOH}}
\end{array}
$$

$$
\begin{array}{ll}
\text{Valine (Val)} & \overset{\displaystyle CH_3\ \ NH_2}{\overset{\displaystyle |\ \ \ \ |}{CH_3.CH—CH—COOH}}
\end{array}
$$

$$
\begin{array}{ll}
\text{Leucine (Leu)} & \overset{\displaystyle CH_3\ \ \ \ \ \ NH_2}{\overset{\displaystyle |\ \ \ \ \ \ \ \ \ \ |}{CH_3CH.CH_2—CH.COOH}}
\end{array}
$$

$$\underset{\substack{| \\ \\ }}{CH_3} \quad \underset{\substack{| \\ \\ }}{NH_2}$$

Isoleucine (Ileu) $CH_3CH_2CH{-}CH.COOH$

$$\underset{\substack{| \\ }}{OH} \quad \underset{\substack{| \\ }}{NH_2}$$

Serine (Ser) $CH_2{-}CH.COOH$

$$\underset{\substack{| \\ }}{OH} \quad \underset{\substack{| \\ }}{NH_2}$$

Threonine (Thr) $CH_3.CH{-}CH.COOH$

(b) Members of this group have two carboxyl groups:

$$\underset{\substack{| \\ }}{COOH} \quad \underset{\substack{| \\ }}{NH_2}$$

Aspartic Acid (Asp) $CH_2{-}{-}{-}CH.COOH$

$$\underset{\substack{| \\ }}{CONH_2} \quad \underset{\substack{| \\ }}{NH_2}$$

Asparagine (Asp—NH$_2$) $CH_2{-}{-}{-}CH.COOH$

$$\underset{\substack{| \\ }}{COOH} \quad \underset{\substack{| \\ }}{NH_2}$$

Glutamic Acid (Glu) $CH_2CH_2{-}CH.COOH$

$$\underset{\substack{| \\ }}{CO.NH_2} \quad \underset{\substack{| \\ }}{NH_2}$$

Glutamine (Glu—NH$_2$) $CH_2CH_2{-}CH.COOH$

(c) Members of this group have two amino groups:

$$\underset{\substack{| \\ }}{NH_2} \qquad\qquad \underset{\substack{| \\ }}{NH_2}$$

Lysine (Lys) $CH_2CH_2CH_2CH_2{-}CH.COOH$

$$\underset{\substack{| \\ }}{NH_2}OH \qquad\qquad \underset{\substack{| \\ }}{NH_2}$$

Hydroxylysine (Hylys) $CH_2CH.CH_2CH_2{-}CHCOOH$

$$\underset{\substack{| \\ }}{NH_2} \qquad \underset{\substack{| \\ }}{NH_2}$$

Arginine (Arg) $CNHCH_2{-}CH.COOH$

$$\underset{\substack{| \\ }}{NH_2}$$

(d) Members of this group contain sulphur

$$\underset{\substack{| \\ }}{SH} \quad \underset{\substack{| \\ }}{NH_2}$$

Cysteine (CySH) $CH_2{-}CH.COOH$

Cystine (CySSCy)

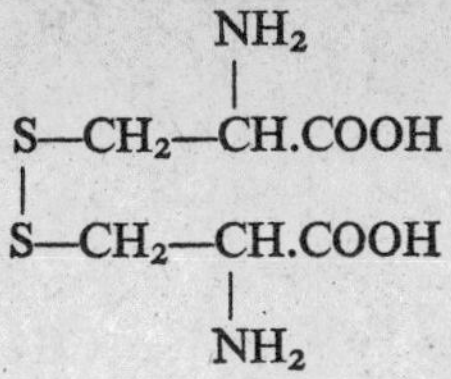

NH2
S—CH2—CH.COOH
S—CH2—CH.COOH
NH2

Methionine (Met)

S—CH3 NH2
CH2CH2—CH.COOH

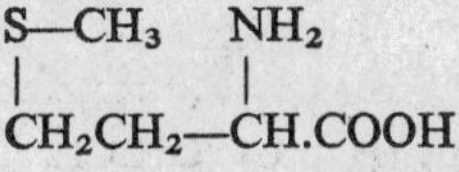

(*e*) Members of this group are aromatic amino-acids:

Phenylalanine (Phe)

NH2
CH2—CH.COOH

Tyrosine (Tyre)

NH2
HO
CH2—CH.COOH

Tryptophan (Try)

NH2
C—CH2—CH.COOH
CH
N
H

Histidine (His)

N—CH NH2
HC C—CH2—CH.COOH
N
H

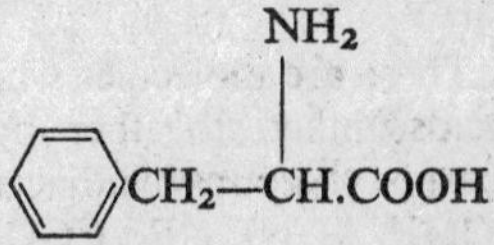

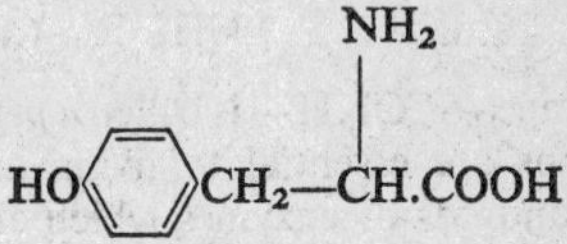

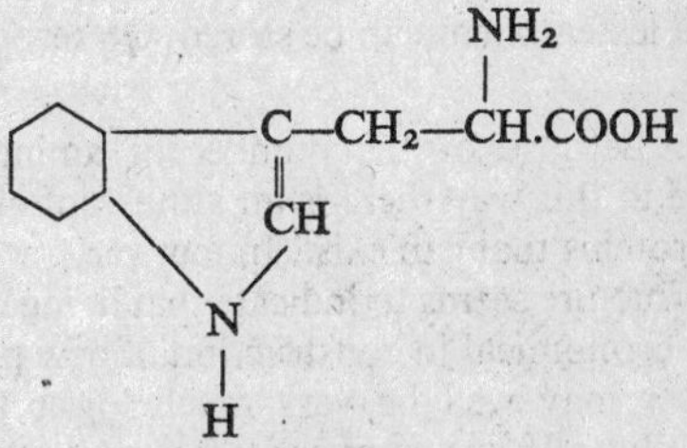

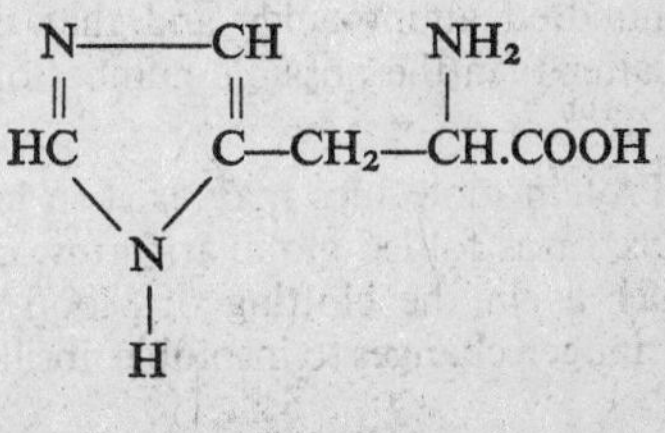

$$CH_2$$

Proline (Pro)

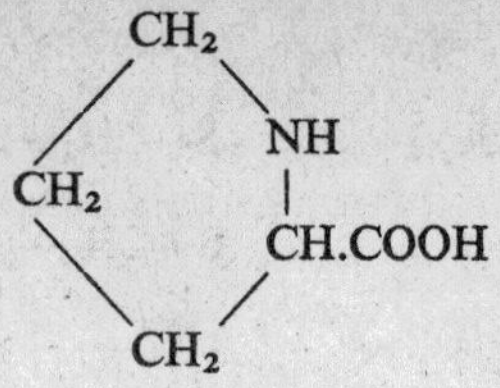

$$CH_2 \qquad\qquad NH$$
$$|$$
$$CH.COOH$$

$$CH_2$$

$$CH_2$$

Hydroxyproline (Hypro) OH.CH

$$NH_2$$
$$|$$
$$CH.COOH$$

$$CH_2$$

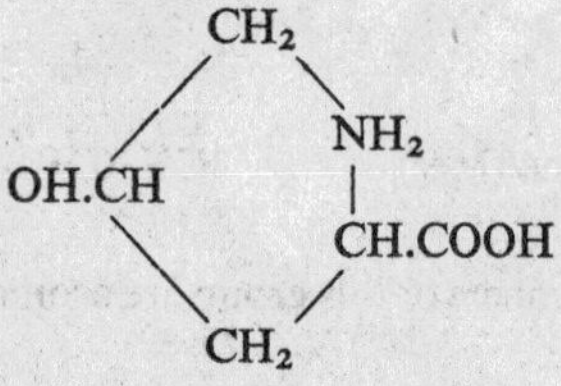

9. Proteins. These are molecules which consist of large numbers of amino-acids linked together by the condensation of the amino and carboxyl groups of adjacent molecules' acids:

$$NH_2 \; R. \; COOH + NH_2 \; R. \; COOH \rightarrow$$

$$NH_2 \; R. \; \boxed{CONH} \; .R. \; COOH + H_2O$$

The linkage —CONH— is called a *peptide linkage*.

Proteins are essential to living organisms as they form the many hundreds of enzymes which allow the varied reactions of metabolism to proceed. They are also used structurally in cells and to a lesser extent can be stored. On respiration proteins yield 17·6 J/g.

NOTE: Because of the number of amino-acids that can be joined in this way there is an almost infinite number of different proteins that can exist. In fact very recent research on protein structure seems to indicate that living organisms have been fairly economical in construction of new proteins so the actual number may well be very much lower than the theoretical possibility. If we suppose that each protein contains one hundred amino-acids and that there are twenty-one of the latter then the possible number of proteins that could exist is 21^{100}.

Protein molecules may exist in long chains, the fibrillar proteins, or as folded globular forms. One form may change to the other as in the clotting of blood where the soluble globular fibrinogen changes to insoluble fibrillar fibrin.

In the last two decades the molecular structure of certain pro-teins such as insulin and haemoglobin has been investigated in great detail. The insulin molecule was found to consist of fifty-one amino-acids of seventeen different types.

On hydrolysis by proteolytic enzymes, the peptide links break and smaller units called polypeptides result. These in turn may be hydrolysed to peptides until tri- and dipeptides with only three or two amino-acids remain. The breaking of the last pep-tide links gives separate amino-acids.

10. Nucleic acids. These are the active constituents of the cell nucleus which allow the coding and transfer of genetic informa-tion and are also found in the cytoplasm associated with *protein synthesis*. Nucleic acids are long chain polymers built up by con-densation of smaller units which in this case are nucleotides. They consist of very many of these units in just the same way as proteins consist of many amino-acids. Like proteins they have very high molecular weights.

Nucleotides are themselves made of various bases in combina-tion with pentose sugar (i.e. a 5-carbon sugar) and phosphoric acid. The bases depend on which sort of nucleic acid is involved and the sugars are either ribose, $C_5H_{10}O_5$ in ribose nucleic acid or deoxyribose, $C_5H_{10}O_4$ in deoxyribose nucleic acid (*see* Fig. 31). The two acids are known as *RNA* and *DNA* respectively.

The bases in RNA are cytosine, uracil, adenine and guanine while those in DNA are cytosine, thymine, adenine and guanine. Both the nucleic acids are long chain polynucleotides with the consecutive sugar linked by phosphate bonds.

NOTE: In 1953 Watson and Crick using *X*-ray crystallography were able to suggest the molecular structure of DNA and RNA. The clearest picture emerges for DNA which is thought to con-sist of a double helix of the polynucleotide strands linked across the adjacent sugars by pairs of bases. In DNA these base pairs are always adenine and thymine or guanine and cytosine. In more recent years the detailed structure of the various types of RNA has been elucidated, including the nature of the binding sites involved in tRNA and mRNA.

While the DNA is located in the nucleus and is the material that carries the genetic code, RNA occurs both in the nucleus and the cytoplasm and is responsible for the actual process of making proteins by the cell (*see* V, **18**). All living organisms

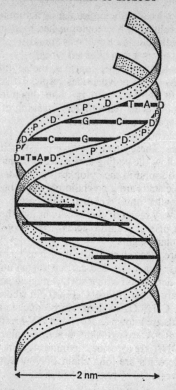

FIG. 31   *The double helix of deoxyribose nucleic acid.*

The base sequences that make up the rungs of the ladder determine the code for protein synthesis. D = deoxyribose sugar. P = phosphate. A = adenine. T = thymine. G = guanine. C = cytosine.

contain both DNA and RNA, except for certain viruses which only have the latter.

A discussion of the role of nucleic acids is found in V, **17–21**.

ENZYMES, VITAMINS AND OTHER CO-FACTORS

11. The properties of enzymes. Additional proteinous substances of cells are *enzymes*, which catalyse reactions and allow them to take place at high speeds at the relatively low temperatures of living organisms. They do this by presenting surfaces on which the reacting molecules can come into close contact and so combine more rapidly. Enzymes lower the activation energy of the substrates on which they act but do not affect the total energy yields of a particular reaction (*see* Fig. 32).

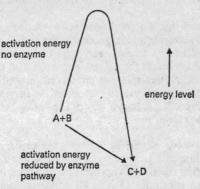

FIG. 32 *One of the most important characteristics of enzymes is the low energy level at which reactions will occur. This allows cell reactions to proceed rapidly at the comparatively low temperatures of living matter.*

Chemically, enzymes consist of a protein part to which is usually bound a co-enzyme, a smaller unit which determines the reaction catalysed by the enzyme. Certain vitamins act as co-enzymes (*see* 19). Besides co-enzymes, inorganic ions of various sorts are often necessary for the enzyme to do its work. These ions are thought to link the enzyme to its substrate during a reaction.

Some of the important properties of enzymes and enzyme-catalysed reactions are as follows:

(*a*) Enzymes tend to be specific for certain substrates or groups of closely related substrates. As the activity of the enzymes is determined by its configuration this will "fit" only the configuration of a certain type of substrate molecule (in the same way that a key fits only one type of lock).

(*b*) The rate of reactions catalysed by enzymes is dependent on temperature, like other chemical reactions. For an increase of

10°C the reaction rate is approximately doubled which may be expressed by stating $Q_{10} = 2$. Unlike other chemical reactions this increase in reaction velocity tails off at much above 30°C as the enzyme itself becomes denatured at higher temperatures.

(c) As the behaviour of proteins and the extent to which they ionise is determined by pH, so this too has an effect on enzyme activity. Particular enzymes tend to act within a narrow range of pH.

(d) Rates of reaction are determined by the concentrations of enzyme and substrate and in general a very small amount of the former will cause reactions in very large quantities of the latter. An important constant for a given enzyme, called the Michaelis constant or K_m, is that substrate concentration which allows half the maximum reaction velocity to occur, the enzyme concentration being fixed. K_m is thus a way of expressing the affinity that an enzyme has for its substrate.

12. Possible mode of enzyme action in the condensation of two substrates. Enzymes tend to be quite specific for the substrates with which they react. If an enzyme (E) is providing the conditions whereby two substrates (A and B) condense together the reaction might take place as in Fig. 33. This is obviously a very

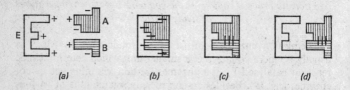

FIG. 33 *Diagram of a condensation reaction involving an enzyme.* (a) E = enzyme, A and B = substrates. (b) A and B come together on the surface of E. (c) A and B combine, releasing, in this case, a molecule of water. (d) The new condensed molecule A+B comes away from the enzyme surface. E is now ready to accept further molecules of A and B.

much simplified scheme as all the molecules involved exist in three dimensions and many have a molecular weight of 10^5 or so. Nevertheless it shows how the configurations are likely to be specific between enzyme and its substrate(s). In such a sequence some of the steps will require energy, that is they will be endothermic. For these steps the cell must provide ATP which yields the necessary chemical energy for the reaction.

13. The classification of enzymes. By the early 1960s more than seven hundred enzymes had been recognised and classified and more are being added to this total each year. While there have been various attempts to standardise the nomenclature and classification of enzymes, the position is still rather flexible and many of the older names are in use. A common method for the naming of an enzyme was to alter the last three letters of the substrate on which it acted and change them to -*ase*. Thus we have the sugar maltose acted on by the enzyme maltase.

Some of the more important classes of enzymes that are widespread in living organisms are (*a*) hydrolases, (*b*) oxidoreductases, (*c*) transferases, (*d*) isomerases and (*e*) additives (*see* **14–18** below).

14. Hydrolases (or hydrolytic enzymes). These bring about the hydrolysis or condensation of substrates by the addition or removal of water:

$$AB + H.OH \underset{\text{condensation}}{\overset{\text{hydrolysis}}{\rightleftarrows}} AH + B.OH$$

They are more familiar as the intestinal juices of animals but they also occur in seeds and other parts of plants. Hydrolases are further sub-divided according to the class of substrates on which they act and are as below:

(*a*) *Proteases* (or proteolytic enzymes) act on proteins or their sub-units such as peptides. They catalyse the breaking of the peptide linkages which join amino-acids —CO.NH— to —COOH and NH_2—. Those proteases such as pepsin and trypsin that work on peptide linkages within the body of the protein are termed endopeptidases while others, the exopeptidases, hydrolyse terminal peptide links. The latter group consists of amino peptidases which attack peptide links next to the NH_2 end of the molecule and carboxypeptidases which hydrolyse the links at the —COOH end. These exopeptidases tend to act on smaller units such as the tri- and dipeptides which themselves result from endopeptidase activity on proteins.

(*b*) *Carbohydrases* bring about the hydrolysis of carbohydrates. The reactions that they catalyse can be summarised in the following table:

TABLE III.   PRINCIPAL REACTIONS OF CARBOHYDROSE ENZYMES

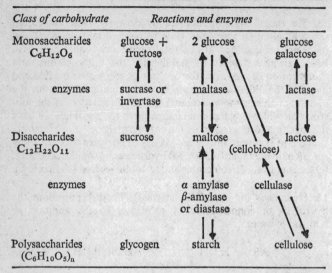

Class of carbohydrate	Reactions and enzymes		
Monosaccharides $C_6H_{12}O_6$	glucose + fructose	2 glucose	glucose galactose
enzymes	sucrase or invertase	maltase	lactase
Disaccharides $C_{12}H_{22}O_{11}$	sucrose	maltose (cellobiose)	lactose
enzymes		α amylase β-amylase or diastase	cellulase
Polysaccharides $(C_6H_{10}O_5)_n$	glycogen	starch	cellulose

NOTE: It should be noted that the amylases that hydrolyse starch consist of two types: α-amylase breaks C to C linkages in both the amylopectin and amylose that make up the starch, while β-amylase breaks only terminal links. A further component of amylase complex breaks the C to C linkages in the amylopectin.

(c) *Lipases* hydrolyse fats into fatty acids and glycerol according to this general equation:

$$
\begin{array}{c}
R_1COOCH_2 \\
| \\
R_2COOCH + 3H_2O \rightleftharpoons \\
| \\
R_3COOCH_2
\end{array}
\quad
\begin{array}{c}
R_1COOH \\
\\
R_2COOH + \\
\\
R_3COOH
\end{array}
\quad
\begin{array}{c}
CH_2OH \\
| \\
CH.OH \\
| \\
CH_2OH
\end{array}
$$

(d) *Nucleases* hydrolyse nucleic acids into the bases, sugars and phospates out of which they are made.

15. Oxidoreductases (oxidases and dehydrogenases). These allow oxidation or reduction reactions to take place. They are essential for both aerobic and anaerobic respiration and their detailed functions are considered under this heading (*see* VIII, 5).

The oxidases, such as the cytochromes, catalyse the addition of oxygen, while the dehydrogenases remove hydrogen, equivalent to oxidation (*see* Fig. 34). The nomenclature of dehydrogenases

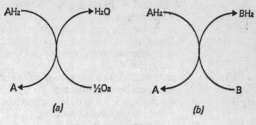

FIG. 34.

(*a*) Action of oxidases. (*b*) Action of dehydrogenases.

and their co-enzymes is in a particularly complex state and is outlined in VIII, 5.

16. Transferases. These catalyse the transfer of a specific group from one substrate to another. Examples are transaminases which carry $-NH_2$ groups and are important in amino-acid synthesis and phosphorylases which add $-PO_4$ to a substrate. The latter are used in certain active transport systems across membranes as well as in the initial stages of respiration.

17. Isomerases. These bring about the change of one isomer into another, as for example that of glucose-6-phosphate into fructose-6-phosphate in anaerobic glycolysis. The enzyme here is phosphohexoisomerase.

18. Additives. These enzymes catalyse the addition of certain particular molecules to others. Thus water is added to fumaric acid to give malic acid in the Krebs cycle. Another example is the addition of CO_2 to acetaldehyde to give pyruvic acid by the enzyme carboxylase.

19. The importance of certain vitamins as co-enzymes. It has been explained above that enzymes consist of a protein part joined to a co-enzyme which is essential in determining its whole activity. The vitamins that really stand out as co-enzymes are those of the B complex (*see* Table IV). When the basic outlines of anaerobic

TABLE IV.   VITAMINS THAT ACT AS CO-ENZYMES

Vitamin	Class(es) of enzyme for which the vitamin acts as a co-enzyme
Nicotinamide:	as nicotinamide - adeninine - dinucleotide (NAD) or its phosphate (NADP) for dehydrogenases.
Riboflavin (B_2):	as co-enzyme for flavoproteins which are oxidases.
Biotin:	as co-enzyme to carboxylase (CO_2 additive) enzymes.
Thiamin (B_1):	as co-enzyme thiamin pyrophosphate for certain oxidases, transferases.
Pyridoxal (B_6):	as co-enzyme for certain transaminases, decarboxylases and isomerases.
Folic acid:	as co-enzyme for transferases.
Cyanocobalamin (B_{12}):	as co-enzyme for isomerases.
Pantothenic acid:	also known as co-enzyme A it is a transferase enzyme by which groups from all types of substrate are introduced into the Krebs cycle.

and aerobic oxidation pathways have been mastered (VIII, **5, 6**) it is worth making a flow diagram and inserting the many steps that are dependent on one or other of the B vitamins. The physiological symptoms that follow deficiency of this group can then be clearly understood.

From Table IV it can be seen just how many enzyme systems of the organism would not operate without the presence of the appropriate vitamin co-enzyme. This knowledge gives us an idea of the role of these vital nutritional substances at the biochemical level.

20. Inorganic ions as co-factors in enzyme function. Certain ions seem necessary for the efficient working of particular enzymes and it has been suggested that they assist in linking together the substrate and enzyme.

Major ion-enzyme associations that have been found by biologists are as follows:

(*a*) Fe^{++}, Cu^{++} and Mo^{++} for certain oxidases;

(*b*) Zn^{++}, Mg^{++} and Mn^{++} for certain hydrolases and additive enzymes; and

(*c*) K^+ and Mn^{++} for certain transferases.

As with the vitamins this does not represent all the organism's requirements and there are many other ions important in its activity or structure which do not act as enzyme co-factors.

PROGRESS TEST 4

1. What do you understand by the study of biochemistry?   **(1)**
2. Name two techniques commonly used by biochemists.   **(2)**
3. Into which major categories can carbohydrates be classed? **(3, 4, 5, 6)**
4. What are furanose and pyranose rings? Why are they important?   **(4, 5, 6)**
5. What are fats?   **(7)**
6. Why are fats commonly stored in living organisms?   **(7)**
7. What are amino-acids? Name four amino-acids.   **(8)**
8. How is it possible to have so many different types of protein?   **(9)**
9. What bases are found in DNA?   **(10)**
10. What are the characteristics of enzymic reactions?   **(11)**
11. Name three different types of enzyme.   **(13)**
12. What relationship exists between enzymes and most of the B group of vitamins?   **(19)**

EXAMINATION QUESTIONS

The figures in **bold** type indicate the marks allocated to each question or part question.

1. (a) The formula of any amino acid may be represented as $NH_2R.COOH$. In the space below show how *three* such amino acid molecules may be condensed by the formation of peptide linkages between them.

(b) Briefly explain how it is possible for the extremely large number of proteins known to occur naturally to be formed from only 22 different amino acids.

(c) Show how a disaccharide such as sucrose may be hydrolysed to yield two monosaccharide molecules.

(*Total* **10**) *part question, Associated Examining Board*, 1976

2. What do you understand by the term lipid? Discuss the importance of lipids in cell membranes and prevention of desiccation.

after Oxford and Cambridge, 1974

3. Review the properties of enzymes, including in your dis-

cussion *two* of the following: (*a*) active sites, (*b*) inhibitors (*c*) co-factors.

Oxford and Cambridge, 1976

4. Discuss the importance in the nutrition of man of (*a*) vitamins, (*b*) mineral salts, and (*c*) fats.

London, 1975

5. What is an enzyme? By reference to named digestive enzymes, give an account of the factors that influence the activity of enzymes. Explain why several enzymes are needed in any metabolic pathway.

London, 1975

6. (*a*) What is meant by (*i*) specificity and (*ii*) denaturation of an enzyme? (**6**)

(*b*) Discuss the ways in which (*i*) temperature and (*ii*) pH may affect the activity of an enzyme. (**10**)

(*c*) How may starch be converted to glucose? Would you expect the same enzymes to be involved in the *synthesis* of starch? Give reasons for your answer. (**4**)

London, 1976

CHAPTER V

The Cell

CELLS AND THEIR STUDY

1. The cell concept. The fact that living things are made up of cells has been clear since the mid-eighteenth century.

Cells are those units of protoplasm which are controlled by a single nucleus and whose boundaries are limited by a cytoplasmic membrane. The typical cell in which an organised nucleus is present is termed eukaryotic. Cell nuclei depend on the process of diffusion to maintain the organisation of their cytoplasm and for this reason the amount of cytoplasm associated with a nucleus is limited. Most cells are of the order of 10μm in diameter and those that are not, e.g. many eggs, are mainly full of inert storage matter.

Most animals and plants have bodies consisting of many thousands of millions of cells. It has been shown that for man the daily turnover of cells is in the region of 10^9. In multi-cellular organisms the cells tend to be highly specialised to carry out particular physiological functions.

2. Levels of study in biology and the size ranges involved. The study of structure can be made at several different levels:

(*a*) *Morphology* is the gross study of the shapes of organisms. The size ranges depend on the organism involved and can be from millimetres to metres.

(*b*) *Anatomy* deals with the structure of organs such as the kidney or the plant leaf. The size ranges tend to be from 0·1 mm upwards. No high-powered microscope is required.

(*c*) *Histology* is the study of tissue structure where the individual cell is the smallest unit so that the size ranges are from 10μm upwards. Optical microscopes are used.

(*d*) *Cytology* is the study of cells and covers all the details that are visible with the light microscope, that is down to $0·2\mu$m in size. Light microscopes are not able to resolve objects below this limit.

(*e*) *Ultrastructure* is the study of cell inclusions and other details of the cell; the sizes range from 0·2–10 μm. This study is only possible with the electron microscope which provides magnifications of up to 10^6. Mitochondrion and chloroplast structure fall into this category.

(*f*) *Molecular study* of structure is the final stage of analysis and may be taken as far as the actual configuration of atoms and larger groups within biological molecules. The dimensions involved are less that 1 nm and the method of investigation is by *X*-ray diffraction.

NOTE: While the knowledge of the apparatus and techniques for the study of ultra and molecular structure are, of course, outside the scope of this book, the knowledge gained by their use is not. Advances in these fields have done much to revolutionise our understanding of structure and function in living organisms.

3. The diversity of cells. The general differences between plants and animals and their cells have been outlined in I. A further consideration of some of the structures characteristic of plant cells follows in Figs. 35 and 36 below. In this section consideration is given to the diversity of structure and function of cells.

(*a*) In *unicellular plants and animals* all the physiological functions are carried out by the single cell and its various organelles. This necessarily imposes severe limits on specialisation in any one direction.

(*b*) In *multicellular forms* the physiological activities of the organism are shared out and a certain number of the cells become specialised to carry out a particular function to the exclusion of others. Thus "physiological division of labour" evolved together with histological differentiation.

The actual degree to which these processes have been taken is variable and lower organisms, such as *Hydra*, may have one type of cell carrying out a number of quite different functions. In the higher animals and plants cell specialisation is very far advanced and man has at least one hundred different types of cell in his body.

One of the first sorts of specialisation that occurred in evolution was the separation of the body of the organism, its *soma*, from its reproductive or germ cells. In simple algae such as *Chlamydomonas* or *Spirogyra* the whole cell must act as one or more gametes. More complex forms, such as *Fucus* and *Volvox*, have quite separate and specialised male and female gametes.

NOTE: At the start of life, all organisms consist of unspecialised cells, usually termed embryonic. By a variety of means which probably involve gene masking, the potentiality of the initial cells becomes limited and directed in a specific way so that it develops into a specialised type. In some animals, e.g. platyhelminthes, a process of de-differentiation can occur in unfavourable conditions so that the specialised cell returns to its embryonic and omnipotential form, specialising again as conditions improve.

While higher organisms usually retain a pool of cells which keep their embryonic potentialities and are capable of replacing others in wound healing, regeneration, metamorphosis and the like, the bulk of the cells are specialised.

(c) *Tissues* and *organs*. Specialised cells are aggregated into *tissues*, collections of similar forms, and the tissues into systems called *organs* which make up the functional parts of the organism itself. Thus in ourselves nerve cells make up nerve tissues which in turn make up, with other tissues, the brain and nervous system. In higher plants photosynthesising, supporting, conducting and protective tissues make up the organ of the plant we call the leaf.

NOTE: Some of the specialised plant and animal tissues that are found will be dealt with in the context of the physiological functions they perform; i.e. the two examples mentioned above, the nervous system and the leaf, are considered in detail in the sections on animal co-ordination and plant nutrition respectively (*see* XI and VI).

A very brief outline of the main lines of cellular specialisations according to function is given in Fig. 35.

4. A brief outline of techniques for the investigation of cells. Cells may be investigated at the histological level by sectioning of the organism or its tissues by a microtome. The sections are then stained with various dyes and as the affinity of these for particular chemicals is known, much information can be gained.

Normal staining techniques require that the organism be fixed and thus dead. The phase and interference microscopes which distinguish between substances of the same opacity but different refractive indices allow observation of the living structure. The use of ultra-violet light is another way of examining structures, such as the chromosomes, in living cells.

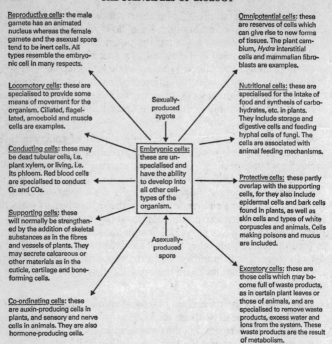

Reproductive cells: the male gamete has an animated nucleus whereas the female gamete and the asexual spore tend to be inert cells. All types resemble the embryonic cell in many respects.

Locomotory cells: these are specialised to provide some means of movement for the organism. Ciliated, flagellated, amoeboid and muscle cells are examples.

Conducting cells: these may be dead tubular cells, i.e. plant xylem, or living, i.e. its phloem. Red blood cells are specialised to conduct O_2 and CO_2.

Supporting cells: these will normally be strengthened by the addition of skeletal substances as in the fibres and vessels of plants. They may secrete calcareous or other materials as in the cuticle, cartilage and bone-forming cells.

Co-ordinating cells: these are auxin-producing cells in plants, and sensory and nerve cells in animals. They are also hormone-producing cells.

Sexually-produced zygote

Embryonic cells: these are unspecialised and have the ability to develop into all other cell-types of the organism.

Asexually-produced spore

Omnipotential cells: these are reserves of cells which can give rise to new forms of tissues. The plant cambium, Hydra interstitial cells and mammalian fibroblasts are examples.

Nutritional cells: these are specialised for the intake of food and synthesis of carbohydrates, etc. in plants. They include storage and digestive cells and feeding hyphal cells of fungi. The cells are associated with animal feeding mechanisms.

Protective cells: these partly overlap with the supporting cells, for they also include epidermal cells and bark cells found in plants, as well as skin cells and types of white corpuscles and animals. Cells making poisons and mucus are included.

Excretory cells: these are those cells which may become full of waste products, as in certain plant leaves or those of animals, and are specialised to remove waste products, excess water and ions from the system. These waste products are the result of metabolism.

FIG. 35   *Differentiation of cells and tissue systems involved in the growth of multicellular organisms* (see also *Fig.* 36).

Very high resolution may be obtained by the electron microscope where the object to be examined, cut at minute thicknesses of only a few nm, is placed in a beam of electrons. The stains used are those which will stop the passage of electrons (e.g. heavy metals), and focusing, equivalent to lenses, is supplied by electromagnets. The picture obtained may be focused on a fluorescent screen or a photographic plate.

Despite the very great advantages of the electron microscope, it should be remembered that it cannot examine living or thick material and considerable care has to be taken in the interpretation of electron microphotographs.

The principle of X-ray diffraction is explained in IV, **2**.

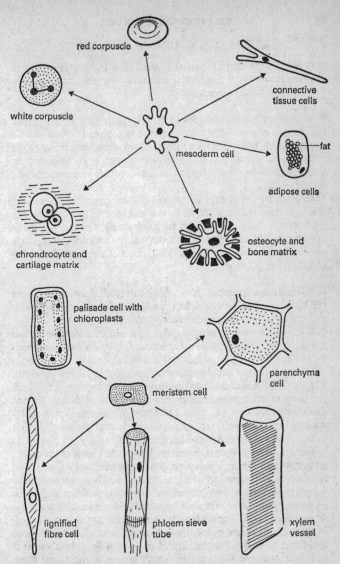

FIG. 36 *A simple scheme showing specialisation of cell types in mammals and flowering plants.*

Only a few examples are given.

ULTRASTRUCTURE OF THE CELL

It should be emphasised that the following are largely descriptive rather than functional accounts; for the latter appropriate references are included.

5. Prokaryotic and eukaryotic cells. It is customary to distinguish two stages in the evolutionary complexity of cells. Very simple organisms such as bacteria and blue-green algae have prokaryotic cells. These are small and generally rather uncomplicated in structure and they lack distinct organelles such as a nucleus and mitochondria.

In all other living organisms the cells are described as eukaryotic and such cells are rather larger and more complex than the prokaryotic type. They tend to have many discrete organelle systems such as mitochondria, a definite nucleus and, in the case of plants, chloroplasts. It is a description of the typical eukaryotic cell that follows.

6. Membranes. Membranes occur not only at the surface of the cell but also *around the nucleus*, the *mitochondria* and the *lysosome*, and are a part of the *endoplasmic reticulum*.

While there were some ideas about membrane structure before the advent of the electron microscope, this instrument has been helpful in providing knowledge of detailed structure.

Cell membranes are normally between 7 and 10 nm in width, and the hypothesis has been put forward that they are very much the same in structure wherever they occur. It also claims that the membrane consists of a double layer of phospholipids sandwiched between two layers of protein. It is thought possible that pores of various sizes are present (*see* Fig. 37).

The above hypothesis is referred to as the "unit membrane" structure and since it was first put forward modifications and alternative structures have been proposed. Thus, according to Lucy (1964) certain membranes have globular micelles or groups of lipids, specifically lecithin–cholesterol, whose tubules are some 4 nm in width and 11 nm in length. In any case it appears that pores, in the sense of actual holes, are only found in nuclear membranes and other so-called "pores" are of protein or fat substance. It is also seen that the structure of membranes is a good deal more complex than initial models suggested. There is also a degree of structural variety to be found in the cell. Some membranes have

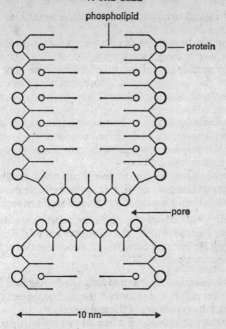

phospholipid

protein

pore

◀——————10 nm——————▶

FIG. 37 *Cell-membrane structure.*

This was first suggested by Danelli on the basis of its behaviour. The veracity of his model was confirmed by the electron microscope. All the membrane systems of the cell have this "unit structure".

a ladder-like appearance; others, like myelin, have high fat and low protein content, and others, like bacterial, have the reverse.

The plasma, endoplasmic and nuclear-membrane systems will be considered separately in the following sections although they may all be closely associated in certain aspects of function.

7. The cell or plasma membrane. The *plasma-membrane* system is that which limits the boundaries of the cell. It has a number of properties:

(*a*) *Function.* Being at the cell surface, the plasma membrane controls the entry and exit of substances into and out of the cell. As we shall see later (*see* VI, **17**), this control may take place by allowing a passive diffusion of substances along their concentration gradients. It may also occur by active uptake, or secretion, against the diffusion gradient.

(b) *Features*. In order to facilitate exchanges with the environment, the plasma membrane is often thrown into numerous folds such as are seen in the micro-villi of the intestine and the brush border of the kidney tubule. Moreover, the membrane may develop actively moving "fingers" and vesicles associated with pinocytosis.

(c) *Associated substances*. Plasma membranes are sometimes covered with layers of mucopolysaccharides which act as templates for the enzymes secreted by the cytoplasm. They also assist control of cell permeability.

(d) *Junctions between membranes*. Specialised fibrillar regions called desmosomes may develop from the cell membrane in conjunction with strengthening effects between cell surfaces.

The individual cells of organisms, especially animals, are held together by surface properties and substances of their cell membranes. Peptide linkages between cell membranes seem important as does the occurrence of Ca^{++}.

8. The endoplasmic reticulum and associated structures. The endoplasmic reticulum is the elaborate membrane system that occupies most of the cytoplasm of the mature cell.

On one side the endoplasmic membranes are partly continuous with the nuclear membranes and on the other they may be contiguous to the plasma membrane (*see* Fig. 38).

(a) *Structure*. On one side of the membranes is the cytoplasmic matrix, a colloidal material containing the soluble proteins and ground substances of the cell, while on the other a large number of spaces or cisternae are enclosed.

(b) *Function*. It is suggested that the endoplasmic membranes and their cisternae have a role in the transport of substances in the cell, in the provision of a large surface area for chemical reactions, in the support and in the localisation of pH and potentials in different parts of the cytoplasm.

Many enzymes, for example those concerned with anaerobic glycolysis, have been associated with the cytoplasmic matrix on one side of the membranes while synthesised material forms within the cisternae.

(c) *Ribosomes*. In this latter context the walls of the endoplasmic membranes have large numbers of RNA-containing particles, called *ribosomes*, situated on them. There may be several

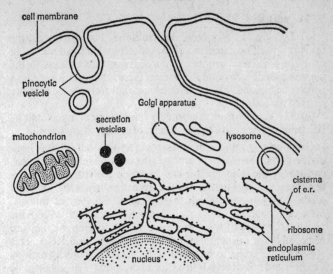

FIG. 38 *Some of the organelle systems of the typical cell as shown by the electron microscope.*

thousands of these in a cell engaged in active synthesis and they are undoubtedly concerned in protein formation by the cytoplasm. In sites of particular activity, numbers of ribosomes are found associated together as polysomes. (The detailed coding of genetic information and its relation to protein synthesis is described in 20.)

(*d*) *Golgi apparatus*. A special development of the endoplasmic membranes is the *Golgi apparatus* which consists of sets of membrane systems with especially large flattened cisternae and with no ribosomes on the walls. The system is discrete in animal cells but dispersed in plants. It is thought that substances synthesised within the cisternae of the endoplasmic reticulum are transported and stored in the larger cisternae of the Golgi apparatus. It must be remembered that the whole system of cisternae and membranes is continuous throughout the cytoplasm. Once collected within the vesicles of the Golgi the secretions of the cells are moved towards the cell membrane where individual packages are "blebbed off". Enzyme- and mucous-producing cells of the gut walls are rich in active Golgi apparati, as are all secretory cells.

9. The nucleus.

(a) *The nuclear membrane* or envelope that surrounds the nucleus is itself an extension of the reticulum described above. In many cases it has a number of pores giving direct access to the nuclear material. Around the nucleus there is a large cisterna into which nucleic acids can be secreted and thus transported to other parts of the cell.

(b) *The nucleus itself* is normally spherical or elliptical and its size tends to be relative to the amount of cytoplasm in the cell.

Cell nuclei go through phases of activity but in the non-dividing state, known as interphase, the protoplasm of the nucleus consists of the fluid nuclear sap in which are suspended many fine threads of chromatin. These chromatin threads are the basis of the chromosomes which thicken up at the start of mitosis. Spheres of nucleic acid called nucleoli are present within the nucleus. The size of these decreases at cell division, as they are responsible for the synthesis of ribosomal structural RNA. This RNA is only going to be made when the cell is actively making proteins and not when it is in a state of division.

10. Mitochondria.

These are spherical or elliptical bodies that are found in the cytoplasm. They may make up as much as one-third of its bulk and there are several thousand in each cell.

(a) *Structure.* The mitochondria are some 0.5μm in width and several μm in length. In section they are seen to be composed of an outer and inner membrane system and a fluid-filled interior. Between the inner and outer membrane is a space or cisterna and while the outer membrane is smooth the inner is thrown into numerous folds called cristae (*see* Fig. 39).

(b) *Function.* As we shall see in VIII, **12**, the whole mitochondrion is concerned with the latter stages of aerobic respiration and the site of ATP generation for the cell. The sites of the enzyme paths involved in this process have been located within the mitochondrion.

(c) *Origin.* Many authorities believe that mitochondria could once have been free living cells of a primitive nature and that such cells might have been engulfed by other cells to give mutually beneficial symbiotic units, paving the way to the evolution of the advanced eukaryotic cell from the simple structured and minute prokaryotic cell.

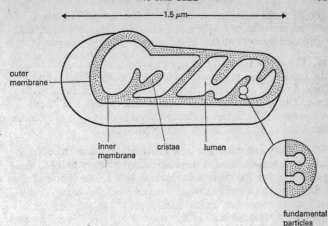

FIG. 39 *A mitochondrion, the "power-house" unit of the cell.*
Sometimes these organelles are more rod shaped and in other cases
they may be spherical

11. Lysosomes. These are inclusions found in cells that vary from
0.2μm to 0.8μm in size. They consist of a membrane enclosing a
fluid-filled vacuole. The contents of the vacuole are a variety of
hydrolases including those which will digest polysaccharides,
proteins and nucleic acids.

It is clear that in normal cells these enzymes are kept insu-
lated from the rest of the protoplasm. Where the lysosomes
break down, due to injury, death or some other cause, self-
digestion or autolysis of the cell occurs.

Lysosomes are of particular importance in cells that engulf
large nutrient particles as the latter may be introduced into the
lysosome and digested there before being made available to the
cell. They are also important in the reorganisation of cells involved in metamorphosis and wound healing.

SPECIAL FEATURES OF THE ULTRA-STRUCTURE
OF PLANT CELLS

12. The plant cell wall. One of the characteristic ways in which
the plant cell differs from that of the animal cell is in the posses-

sion of a cell wall outside its plasma membrane. The plant cell wall is composed of (*a*) the middle lamella; (*b*) primary cell wall; (*c*) secondary cell wall (*see* **13–15** below).

13. The middle lamella. Actively dividing cells are to be found in the meristematic regions of plants and as these cells complete their division a plate of droplets forms between the daughter cells. This is the beginning of the middle lamella and later it is reinforced with calcium and magnesium pectate. It is the middle lamella which holds together neighbouring cells.

14. Primary cell wall. The new plant cell continues to differentiate and grow and as it does so cellulose is synthesised, and, passing across the plasma membrane, lays a foundation of the primary cell wall (*see* Fig. 40).

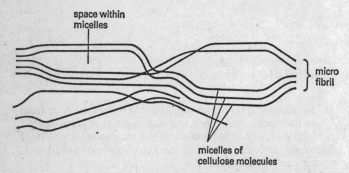

space within micelles

micro fibril

micelles of cellulose molecules

FIG. 40 *Organisation of the cellulose of the primary cell wall.*

This cellulose is formed in bands called microfibrils and magnifications available with the electron microscope show that these are themselves composed of micelles. The latter are strands of cellulose in a crystalline state and some hundreds of nm long.

Between the micelles are spaces and these may be filled with other substances such as hemicelluloses, pectates, lignin, etc. The distinguishing feature of the initial or primary cell wall is its plasticity allowing the growth of the cell to continue. For this reason the addition of substances like lignin is small.

NOTE: Micelles are added to the primary wall from the inside and at first they are orientated at right angles to the cell axis with small inter-micellar spaces. As they are pushed out by the expansion of the cell their orientation changes and their spaces increase. (The change may be likened to lazy-tongs in their shut and open positions.)

Once the cell has reached its mature size the primary cell wall is finished and it provides the cell with a semi-elastic permeable and supporting layer.

15. Secondary cell wall. Many plant cells, however, secrete a further secondary cell wall beneath the primary and this will have highly organised micelles whose spaces are impregnated with lignin (in sclerenchyma or woody tissue) or with extra cellulose (in collenchyma). The secondary cell wall tends to be rigid and impermeable and this characteristic brings about the death of the cell by cutting off essential nutrients from the cytoplasm. In the case of lignified tissues it provides a great deal of extra strength for the support of the plant (*see* Fig. 41).

Other variations in cell walls are found in the cuticularised cells of the epidermis. Here a layer of cellulose and cutin is passed out through the primary wall and middle lamella to form a surface coating to the outer part of the cell. A further layer of pure cutin and wax is added and thus an impermeable surface obtained.

In the suberised cells found in bark a layer of the fatty substance suberin is laid down between primary and secondary walls. If this goes all round the cell it will bring about its death. The cork or phellem cells formed by the cork cambium or phellogen prevent the passage of water and gases into and out of the woody plant.

It should be noted that strands of protoplasm pass from one cell to another, at least through the primary walls, and these maintain the protoplasmic continuity of the plant. Once the secondary wall is formed the strands are severed. These strands are sometimes called plasmodesmata.

16. Chloroplasts. These are derived from colourless leucoplasts called proplastids and form in many plant cells in the presence of light. In higher plants individual chloroplasts are some 5μm by 3μm. Within the envelope surrounding the chloroplast a complex system of tube-like lamellae are found. In certain places along their length the lamellae are concentrated into regions called grana. Each granum has been likened to a pile of pennies, owing to the discoid shape, and the numbers found from one species to

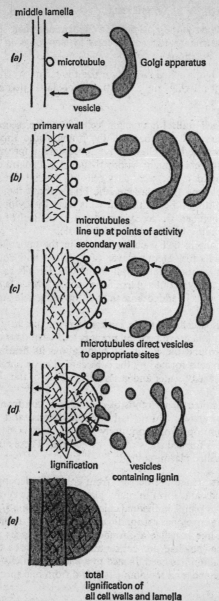

(a) middle lamella
microtubule
Golgi apparatus
vesicle

(b) primary wall
microtubules
line up at points of activity
secondary wall

(c) microtubules direct vesicles
to appropriate sites

(d) lignification
vesicles
containing lignin

(e) total
lignification of
all cell walls and lamella

FIG. 41

another are very variable. In the chloroplasts of spinach there are between forty and sixty grana per chloroplast (*see* Fig. 42).

Surrounding the grana and lamellae is a fluid called the stroma. The chlorophyll is concentrated on the grana while the enzymes involved in the dark reaction part of photosynthesis (*see* VI, **26**) are to be found in the stroma. Starch grains, the products of photosynthesis, form in the stroma, which may also contain oil droplets. Both chloroplasts and mitochondria (*see* **10**) are now known to contain their own DNA and RNA, and this is sufficient to allow for coding for at least some of their own protein structures independently of nuclear DNA.

PROTEIN SYNTHESIS IN THE CELL

17. Organisation of genetic information in the nucleus. The chromosomes are made up of *deoxyribose nucleic acid* or DNA whose structure has been outlined in II. It will be recalled that DNA consists of a double helix of nucleotides with the base pairs adenine and thymine, and guanine and cytosine.

(*a*) *The gene.* In the nucleus there is a complete chemical code for protein synthesis in the cell. The DNA helices are themselves organised into tight spirals which can be seen as actual chromosomes. Within a single nucleus there may be many yards of DNA but only in certain places along these strings are there meaningful code patterns for synthesis. These specific areas are termed *genes*

FIG. 41 *Process of lignification of cell wall of xylem vessels.*

(*a*) Newly divided cell. The middle lamella is made of a pectic substance based on galactose polymers, bound with calcium. The Golgi body buds off parcels of vesicles containing more pectates, together with cellulose. (*b*) Formation of primary wall. The primary wall is made largely of cellulose and hemicellulose, with some pectates. The Golgi body continues to bud off vesicles of cellulose. (*c*) Formation of secondary wall. The secondary wall has the same composition as the primary wall. (*d*) Process of lignification starts. The Golgi body now produces vesicles containing the aromatic polymer lignin, and this coats the micelle network of the cell walls and middle lamella. (*e*) Lignification complete. The cell walls and lamella are now totally lignified. The final composition of this complex is: pectates, 2%; celluloses, 70%; lignin, 28%. The Golgi body and the rest of the cytoplasm are now dead and shrivelled up, leaving a lumen (space) for water conduction through the centre of the xylem vessel.

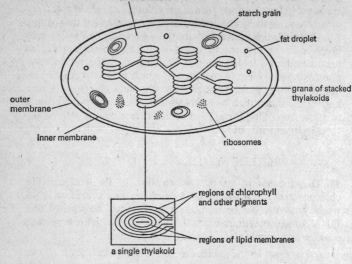

FIG. 42 *A chloroplast*.

The photosynthetic pigments are stacked in the coin-shaped grana while the dark stage reactions take place in the surrounding stroma.

and between them are probably blank areas. It has been observed that in certain regions of chromosomes at certain times "puffs" appear which may correspond to intense gene activity.

NOTE: In some insect larvae, e.g. *Chironomus*, injection of the metamorphosis hormone *ecdysone* causes puffs to appear in regions of the chromosomes within a short time (*see* XI, **20**). This is taken to show that hormones can act by switching on certain gene regions in their target cells and thus causing changes in protein synthesis such as occur during metamorphosis.

(*b*) *Synthesis of template (or messenger) RNA by the gene.* Bases present in the nuclear plasm, together with ribose sugars and phosphates come together at the region of a gene for synthesis of an RNA molecule. For this to take place the double helix of the DNA unwinds and the RNA constituents line up so that the bases are opposite their partners, i.e. adenine opposite uracil (which replaces thymine in RNA) and guanine opposite cytosine (*see* Fig. 43). Condensation of the RNA strands is brought about by the enzyme RNA polymerase. The template RNA molecule comes off the sites on the DNA and makes its way out through the nuclear pores to the ribosome of the endo-

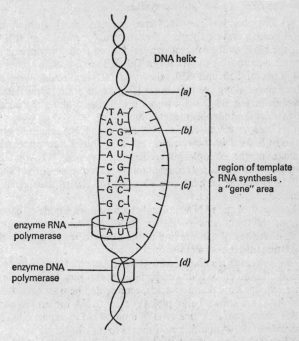

FIG. 43 *The synthesis of a strand of template RNA on the DNA of the gene.*

(*a*) Helix unwinds. (*b*) Active strand of DNA with base template. (*c*) Synthesis of RNA with base pattern determined by code of DNA strand. (*d*) Helix rewinds. Later the RNA strand will travel out from the nucleus to a site on the ribosomes.

plasmic reticulum. The DNA helix rewinds on the enzyme DNA polymerase. Some cell organelles such as the mitochondria and chloroplasts have their own DNA and can replicate themselves in the cytoplasm.

NOTE: Later in this chapter we shall consider how the gene code is passed on from one cell to another first by replication of the chromosome DNA and then by its exact division between two daughter cells.

18. The ribosome and protein synthesis.

(a) *The ribosome.* The ribosomes are made partly from structural protein and partly from the template (or messenger) RNA. All the parts of the ribosome were derived from the nucleus at one time or another in evolution. The structural protein comes in sub-units of S50 and S30 related to their molecular weights. These move along the messenger RNA as its code is translated. As seen in Fig. 38 the ribosomes are arranged along the inside surfaces of the endoplasmic reticulum.

(b) *Transfer RNA—tRNA.* This is soluble RNA, again originally synthesised in the nucleus, but now found free within the vesicles of the endoplasmic reticulum. Transfer RNA is the molecule that will carry specific amino-acids to their correct sites on the ribosome templates so that peptide chains and proteins can be made.

Each end of the tRNA molecule incorporates a code of bases. At one end the code is specific for one of the twenty or so amino-acids that the cell will use and at the other there is a triple code for fitting into the correct site on the ribosome.

(c) *Arranging the amino-acids.* The sequence of events in protein synthesis can be followed by reference to Fig. 44. In the first phase the individual amino-acids, represented here as AA_1, AA_2 and AA_3, are shown being picked up by tRNA molecules, each RNA picking up only one type of amino-acid as explained. The amino-acid–RNA complex now moves to the ribosome.

The second phase shows the transfer RNA becoming arranged along the ribosome according to the fit of triplets of bases (which is why it is called a *triplet code*). Linking of the transfer to the template RNA is enzymically controlled and requires energy.

The third phase shows how neighbouring amino-acids link up by condensation (again requiring energy), forming part of a peptide chain which is subsequently released from the ribosome. Meanwhile the empty tRNA returns to the endoplasmic fluid where it can once again pick up new amino-acids.

It now seems fairly certain that an amino-acid carried by one tRNA molecule is passed to a peptide chain hanging from the ribosome by another tRNA molecule and that only one of a pair

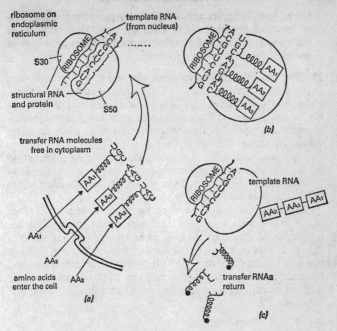

FIG. 44   *An illustration of the successive stages in the synthesis of a peptide chain on a ribosome.*

(*a*) Each different amino acid entering the cell is picked up by a different type of transfer RNA (tRNA). At the opposing end of the tRNA molecule are 3 bases which will correspond to 3 base positions on the ribosomal (template) RNA. (*b*) The tRNA molecules take up specific positions in the ribosome according to the base coding on the template RNA. (*c*) The amino acids combine to form a peptide chain and are released from the ribosome, leaving the tRNA molecules free to return to the cytoplasm. In fact the growing peptide chain is passed backwards and forwards between two sites on the ribosome as each new amino acid is added.

of tRNA molecules is freed at any one time. This is a more accurate picture but in no way changes the basic coding mechanisms described.

Up to two amino-acids are coded and built into a peptide chain in one second.

NOTE: Antibiotics act against particular points in the synthesis of bacterial proteins. Thus some prevent DNA replication, others such as tetracyclin and actinomycin act as inhibitors of protein formation on the ribosomes.

19. The relationship between genes and enzymes. The code of a single gene as fixed into its DNA molecules is passed on, as seen above, to the ribosome. Each gene in the nucleus will be responsible for fixing the template pattern on a great many separate ribosomes. Each type of protein synthesis thus takes place on many similar ribosomes throughout the cell.

It has been established that each gene encodes the information for the synthesis of a single enzyme and it is the enzymes and the reactions that they can catalyse which are responsible for the whole nature of the cell and the organism.

Biochemical and genetic work on certain fungi and other organisms has demonstrated that where particular genes are missing or altered then the cells cannot carry out various syntheses. In fact they lack particular enzymes. A nucleus that has two contrasting genes one of which produces normal protein and the other not will send out both messages to the cytoplasm. Both types of protein form on the ribosomes but only the correct one functions.

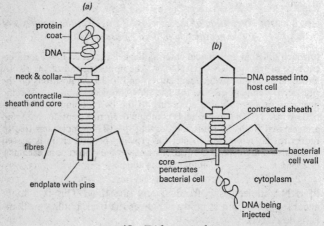

FIG. 45　*T4 bacteriophage.*
(*a*) Free 'phage. (*b*) 'Phage attached to a bacterial cell.

20. Protein synthesis and the triplet code. The above sections have described the basic methods of protein synthesis in the cell and the role of the nucleus and the cytoplasm. It may be asked why a code using three bases was selected rather than one with a greater or lesser number.

If there are four bases and they exist in triplets, then the possible combinations are $4 \times 4 \times 4 = 64$. There are only twenty amino-acids to a code, so a code using triplets of bases is adequate. It is also true that many of these combinations would overlap, i.e. CGA and GAT. If we select non-overlapping combinations, the sixty-four possibilities are reduced to the number that we need. In fact there are a surplus number of combinations and some amino-acids will fit in several triplets. Others are used as full stops between one gene and another.

To prove that genetic information was passed by a triplet code, bases were inserted or deleted from nucleic-acid chains and the resulting peptides analysed.

When a single base or two was either inserted or deleted, the peptide chain beyond the interference read nonsense as compared with the original. When a third member was either deleted or inserted it once again made sense. This can be understood by reference to a chain of three-letter words:

The man saw the dog.
Insert one letter (a):
The man $a$sa wth edo g.
or two (a and b):
The man abs awt hed og.
but insert three (a and b and c):
The man abc saw the dog.

Deletions of letters would work in the same way.

21. The use of bacteriophages in genetic research. In order to make the insertions or deletions within the gene itself, mentioned in **20** above, use was made of the bacteriophage (*see* III, **14**). These are viruses that are parasitic on bacteria and whose structure consists of a protein coat containing DNA and an injection apparatus whereby this genetic material can be passed into the host bacterial cell (*see* Fig. 45). The series of 'phages used were the T2 and T4 types which infect the *Escherichia coli* bacteria commonly found in the gut of mammals.

Essentially the 'phage would pick up a mutant region of DNA from one colony of bacteria and pass it on to another. The progeny of the bacteria then "infected" with the mutant could be picked out readily by spotting the characteristic colonies they make when subsequently cultivated. Such work involved vast numbers of individual cells and the results were available within a matter of hours: thus the bacteria–'phage combination has great advantages in genetic research as compared with the so-long-used *Drosophila*.

As explained in **20** above, one, two or three deletions or insertions of base pairs to a portion of bacterial DNA, through the agency of the 'phage, showed that sense proteins (or enzymes) were made when a triplet had been completely exchanged.

22. Coding the amino-acids. Since the original discovery of the triplet code system for amino-acids on DNA and on messenger and tRNA the actual triplets for individual acids have been established. This work was made possible by the synthesis of various polyribonucleotides, the first actually made being

TABLE V.   AMINO ACIDS AND THEIR RNA BASE TRIPLET CODES

Amino-acid	Code on the ribosome
Alanine	CCG
Arginine	CGC, AGA
Asparagine	ACA, AUA
Aspartic acid	GUA
Cysteine	UUG
Glutamic acid	GAA
Glycine	UGG, AGG
Histidine	ACC
Isoleucine	UAU, UAA
Leucine	UUG, UUC, UUA
Lysine	AAA
Methionine	UGA
Phenylalanine	UUU
Proline	CCC
Serine	UCU, UCC, UCG
Threonine	CAC, CAA
Tryptophan	GGU
Tyrosine	AUU
Valine	UGU

Other possible triplets act as full stops between peptides.

U-U-U. (It should be remembered that all RNAs use the base uracil (U) instead of thymine (T).) This synthetic polymer would latch on to the ribosomes like a natural piece of tRNA and produced a series of phenylalanines (*see* Table V) from the ribosome. Other synthetic polymers were made and the complete triplet code for all the amino-acids pieced together.

23. Jacob–Monod hypothesis. As we shall see in XI, **17**, hormones may work, at least in some cases, by "switching on" particular genes. It is also known that the presence of a substrate appears to cause cells to manufacture the specific enzyme necessary to metabolise the substrate. This phenomenon is termed *enzyme induction*.

It is clear that most of the thousands of genes in a given cell are not working most of the time, and the notion that they could be switched off or on as might be appropriate is an attractive one.

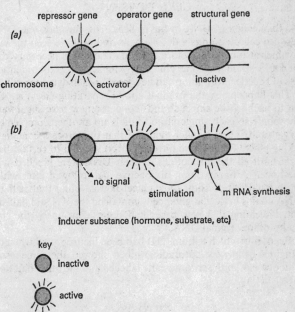

FIG. 46 *Illustration of Jacob–Monod hypothesis.*
(a) Normal inactive state. (b) "Switching on" process.

Some experimental evidence to this effect led to the hypothesis of *gene activation* put forward by two French biologists, Jacob and Monod, in the late 1950s.

They worked on enzyme production in bacteria and as a result produced the hypothesis shown in Fig. 46. The *structural gene* is the one that codes for the enzyme (or some other protein). It is switched off because it only works when stimulated by the adjacent *operator gene*. This in turn is deactivated by the *repressor gene*. When the *activator* (inducer) *substance* is present it combines with, or in some way inhibits, the repressor gene, switching it off. The operator gene is then no longer prevented from working, and it in turn stimulates the structural gene to start coding for mRNA. Protein synthesis will then follow in the cell. This hypothesis now has a good deal of evidence to support it.

CELL DIVISION

24. Replication of DNA in the nucleus. The chromosome DNA between cell division is a double helix and we have seen the way in which it functions. Before the cell divides it is necessary that the DNA strands become completely replicated so that identical sets of genetic information can be passed to each daughter cell.

This happens by the original strand separating (*see* Fig. 47). On to each of the single strands condense new base sugars and phosphate so each single strand builds up another complementary strand. There is much evidence that this process is actually taking place. Radioactive nitrogen (^{15}N) was fed to certain bacteria until all the nucleotides of their DNA were labelled with the tracer. A normal, non-radioactive diet followed after which an internal cell division took place. It was found that half the DNA chains in the daughter cells contained the ^{15}N and half did not, indicating that new bases had been taken up by the split chains of the original DNA.

Beans into which tritium (^3H) had been incorporated when the daughter cells were autoradiographed showed the same result, with one half of the chromosome labelled and the other half non-labelled.

MITOSIS

25. Definition. Normal somatic division of the cell is by *mitosis* which produces *two identical daughter cells*. The process takes

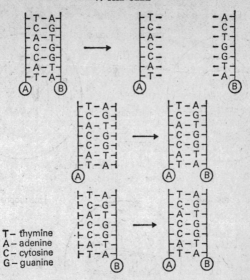

T — thymine
A — adenine
C — cytosine
G — guanine

FIG. 47 *Replication of DNA within the nucleus occurs during cell division.*

The method shown has been confirmed using cells fed with labelled nitrogen.

place in four stages, *prophase, metaphase, anaphase* and *telophase* (*see* **26–28** below) and of these the first and last are comparatively lengthy stages while the second and third take place rapidly (*see* Fig. 48).

26. Prophase. The interphase chromosomes are hydrated and can only be seen by use of special techniques. During prophase the following events take place:

(*a*) The chromosomes appear, i.e. they can be identified with appropriate staining techniques as long coiled threads. They can be seen to be double structures consisting of *two chromatids* joined at a constricted region termed the *centromere*.

(*b*) The chromosomes shorten, by as much as twenty-five times, and thicken. Each chromosome is quite separate (*cf.* meiosis, **31**) and they begin to move out to the periphery of the nucleus. The nucleoleus decreases in size as it is concerned with

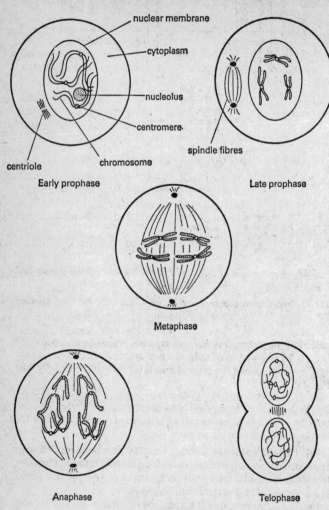

FIG. 48   *Normal cell division by mitosis.*

the synthesis of ribosomal mRNA, which does not occur at this time.

(c) The nuclear membrane begins to disintegrate.

(d) The *centriole*, a small body found just outside the nucleus, divides in half and each half migrates away from the other until the two are found at opposite sides of the nucleus. As the daughter centrioles move apart they give rise to the spindle fibres.

NOTE: The centriole is probably homologous with the basal granule of the cilium. It has the characteristic arrangement of nine fibrils although these may be found sub-divided into smaller groups. It is typical of animal rather than plant cells.

27. Metaphase. The following events are the stages of metaphase:

(a) The chromosomes reach the equator of the cell and become attached to the spindle fibres at the region of the centromere, the nuclear membrane having completely disintegrated.

(b) The centromeres split and the two halves initiate the formation of centromere fibres.

(c) A process of extension of the centromere fibres takes place while at the same time the spindle fibres contract. The net effect of these two activities is that the daughter chromatids of each chromosome begin to move apart starting at the region of the centromere.

28. Anaphase. This takes place rapidly and is a continuation of the drawing apart of the sets of chromatids that began at metaphase. The two sets move towards opposite poles of the cell.

29. Telophase. This is more or less a reversal of the events taking place in prophase:

(a) The divided chromosomes reach their respective poles and begin to hydrate. A nucleoleus is formed.

(b) The nuclear membrane is reconstituted.

(c) The spindle fibres de-differentiate.

(d) The cell divides probably as the result of changes taking place in the cortex and membrane. In plant cells a middle lamella and cell wall is laid down (*see* **13–15**).

MEIOSIS

30. Definition. Meiosis is the form of division which brings about a halving of the chromosome number as well as an interchange of genetic material between homologous chromosomes. In animals meiosis takes place at the formation of the gametes (*see* **39** below) and in diploid plants at the formation of spores (*see* **40** below).

Eight phases can be recognised corresponding to a double mitosis with only one division of the chromosomes (*see* Fig. 49). The important events take place in the prophase I.

31. Prophase I. This phase has several sub-divisions:

(*a*) *Leptotene.* The chromosomes become fixable and elongated; there are fine threads arranged somewhat in the manner of a bouquet of flowers. It is not certain whether the chromosomes are single or double at this stage but most authorities seem to think that the former is the case. Homologous chromosomes are separate.

(*b*) *Zygotene.* Homologous chromosomes come together at first at their tips then later along their whole length. The pairing is very close so that the equivalent region of each member of the pair, or bivalent, is brought into contact.

(*c*) *Pachytene.* The homologous chromosome pairs thicken up and each member can now be seen to consist of two chromatids. The two centromeres are closely opposed.

(*d*) *Diplotene.* The homologous chromosomes now lose their attraction for each other and begin to draw apart except at certain points where one member (or sometimes both) of one chromatid pair has become joined to the other. These points of junction are called *chiasmata* and there are commonly between two and four of them to each bivalent. As may be seen in Fig. 50, part of a chromatid derived from one homologous chromosome has become joined to one from another. This process of *crossing over* is extremely important as it represents the interchange of genetic material between the chromosomes which leads to variation in the offspring.

(*e*) *Diakinesis.* This stage sees a further moving apart of the bivalent pair.

32. Metaphase I. This sees the movement of the chromosome pairs to the equator of the cell. The two centromeres of each pair become attached to spindle fibres from the split centrioles (as in mitosis). The same processes of expansion of centromere fibres and contraction of spindle fibres that take place in mitosis also

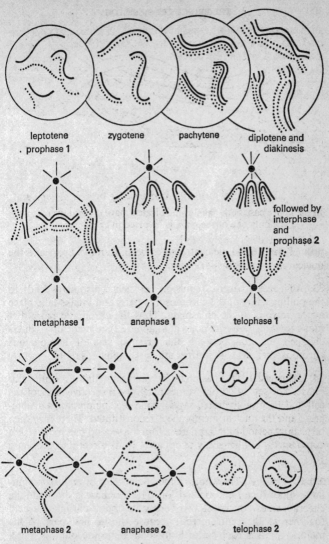

leptotene prophase 1   zygotene   pachytene   diplotene and diakinesis

followed by interphase and prophase 2

metaphase 1   anaphase 1   telophase 1

metaphase 2   anaphase 2   telophase 2

FIG. 49 *Reduction division which normally occurs at gamete or spore formation.*

This is called meiosis.

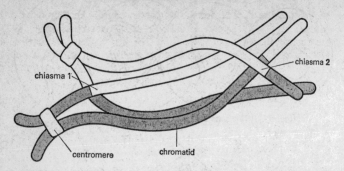

FIG. 50 *Chiasma formation and crossing over at diplotene.*

In the above case both pairs of the bivalent have crossed over; usually
only one of the pairs is involved in exchanges.

occur here. The whole centromeres begin to move apart and the
nuclear membrane disintegrates.

33. Anaphase I. Active repulsion between centromeres leads to
the pulling apart of the chromosome pairs and breaks take place
at the regions of the chiasmata. Thus the chromatid pairs that
move towards opposite poles at anaphase I are made up of inter-
changed material between the original sets of homologous
chromosomes. As with normal mitosis anaphase is rapid.

34. Telophase I. This has much in common with the mitotic telophase. The sets of chromosomes reach the respective poles of the
cell and become hydrated. Meanwhile the spindle fibres disintegrate and the nuclear membrane is reconstituted. There may then
be a short period of interphase before the second meiotic division
commences, but generally telophase I leads directly to prophase
II.

35. Prophase II. The whole of the second part of meiosis is similar to an ordinary mitotic division. The prophase is short and the
chromosomes consist of two widely diverging chromatids held
together at the centromeres. Because of the first part of the
meiotic division each set of homologous chromosomes is represented at this stage by only a single chromosome which contains
at least some material derived from each member of the original
pair.

36. Metaphase II. The spindle fibres very rapidly become reconstituted and the prophase II chromosomes become attached in the equatorial region of the cell by their centromeres. In pollen tetrads the four daughter cells stay together and the plane of division between the two parts of meiosis is at right angles. Each centromere splits and moves away from the other.

37. Anaphase II. Separation of the chromatids takes place.

38. Telophase II. The two sets of chromatids move to opposite ends of the cell. There they become hydrated and the nuclear membrane is reconstituted. Division of the cell takes place.

39. Gametogenesis. In animals meiosis takes place at the formation of gametes. A scheme showing the development of male and female sex cells in animals is as in Table VI.

TABLE VI.   COMPARISON OF SPERMATOGENESIS AND OOGENESIS

Spermatogenesis	*Oogenesis*	
embryonic tissue 2n	embryonic tissue 2n	
↓	↓	
spermatogonia 2n	oogonia 2n	
↓	↓	
primary spermatocyte 2n	primary oocyte 2n	
↓	↓ ↘	
(2) secondary spermatocyte n	secondary oocyte n	first polar body
↓	↙ ↘	
(4) spermatids n	ovum n	second polar body
↓		
(4) spermatozoa n		

More details of the gametes are found in X, 15–18.

40. Meiosis in plants. There seems to be a definite biological advantage in being diploid. This may be due partly to the masking of unfavourable genes by dominant alleles but must also be because of the great increase in heterozygosity and thus phenotypic variation that it allows. The life cycles of various plants are outlined in III and X and by reference to X, **6**, it can be seen how the higher plants have evolved a dominant diploid sporophyte. Meiosis takes place at spore formation and in higher plants microspores produce a male gametophyte and male gametes while megaspores produce a female gametophyte and egg.

PROGRESS TEST 5

1. Distinguish between morphology, anatomy and histology. **(2)**

2. What is meant by a "differentiated cell"? **(3)**

3. What is an electron microscope? **(4)**

4. What is the structure of biological membranes? **(6)**

5. How would you recognise a mitochondrion in an electron microphotograph of a cell? **(10)**

6. How is the cell wall of plants composed? **(12)**

7. What is known of the structure of chloroplasts? **(16)**

8. What is a gene? **(17)**

9. Distinguish between template (messenger) and transfer RNA. **(17, 18)**

10. What is the "one gene, one enzyme" hypothesis? **(19)**

11. How do DNA molecules replicate? **(24)**

12. What sort of cells divide by mitosis? **(25)**

13. Which four phases can be recognised in mitosis? **(25–29)**

14. What sort of cells divide by meiosis? **(30)**

15. Outline the main events of the first prophase of meiosis. **(31)**

16. How may the stages of gametogenesis be related to stages of meiosis? **(39)**

EXAMINATION QUESTIONS

The figures in **bold** type indicate the marks allocated to each question or part question.

1. The nuclei of rat liver cells are composed of protein, DNA, RNA and lipid. Comment briefly on the chemical differences between these compounds. **(7)** What is the role of each of these compounds in the nucleus? **(11)**

(Total **18***) Cambridge*, 1977

2. Show by means of a diagram the structure of deoxyribose nucleic acid (DNA). **(6)**

Discuss the significance of the structure of DNA in relation to:

(a) gene replication (6)

(b) protein synthesis. (6)

(*Total* 18) *Cambridge*, 1977

3. What are the functions of *proteins* in living organisms? (6) Explain, with reference to suitable examples, how the shapes of the molecules may be related to their functions. (12)

(*Total* 18) *Cambridge specimen paper*, 1976

4. Attempt to explain, in language which could be understood by an intelligent person without biological training or knowledge, how the behaviour of chromosomes at meiosis and at fertilization results in variation among offspring. Comment on the usefulness to the species of this variation.

(*Total* 20) *Associated Examining Board*, 1976

5. Use of the electron microscope has made it possible to observe the detailed structure of organelles within the cells of animals and plants. Five kinds of such organelles which can be found in the majority of animal cells are listed below:

Mitochondria, Golgi apparatus, Centrioles, Endoplasmic reticulum with ribosomes, Nuclear envelope.

Choose *four* of these and describe the structure and functions of each, proceeding as follows.

(a) Name it.

(b) Draw a large diagram to show the essential features of its structure. Label this diagram.

(c) Relate its structure to what is known of its activities in the living cell.

(*Total* 25) *Associated Examining Board*, 1974

6. *Read the following account of some of the early stages in the unravelling of the genetic code.*

"From the time of Hammerling's work in the 1920s it was clear that the cell nucleus carried at least the bulk of the hereditary information. Soon after the 1939–45 war an analysis was made of the cell transforming chemical isolated from capsule-forming strains of *Pneumococcus* bacteria. Everything indicated that the active principle was a nucleic acid and more precisely that it was deoxyribonucleic acid (DNA).

"After considerable difficulties an X-ray diffraction photograph was taken of this substance and in 1953 Crick and Watson put forward their interpretation of the molecule as a double helix of phosphates and suagars joined by pairs of pyrimidines and purines. These could be only linked by thymine joining with adenine and cytosine joining with guanine.

"At the time of their first note to *Nature* Crick and Watson said that it had not escaped their notice that the model proposed for DNA would be ideal for coding and replication of hereditary material."

Continue the above account to the stage reached today in molecular genetics. Write briefly but clearly. You will not be penalized for incorrect sequence of subsequent discoveries.

(Total **20***) Associated Examining Board*, 1974

7. DNA is mainly confined to the nucleus, whereas proteins are synthesised in the cytoplasm. Yet DNA is said to control protein synthesis. How can you explain this? **(12)**

What is the evidence for your explanation? **(8)**

after Oxford and Cambridge, 1977

8. Give an account of meiosis. **(8)** How does it differ from mitosis? **(6)** What is the significance of meiosis? **(6)**

Oxford and Cambridge, 1977

9. With the aid of labelled drawings describe the structure of (*a*) a mitochondrion and (*b*) a chloroplast. **(12)** Compare the functions of these two organelles. **(8)**

Oxford and Cambridge, 1977

10. What do you understand by the "cell doctrine"?

Discuss its implications.

London, 1975

11. Write concise notes on *each* of the following: (*a*) mitochondria, (*b*) nucleoli, (*c*) endoplasmic reticulum.

How has electron microscopy changed our views on the structure of cells? (*See also* Chapters III and VI.)

London, 1975

12. What is the *evidence* that deoxyribonucleic acid (DNA) is involved in the mechanism of heredity? Briefly describe the structure of DNA. How do the structure and the role of DNA differ from those of ribonucleic acid (RNA)? **(9, 6, 5)**

London, 1976

PART TWO

COMPARATIVE PHYSIOLOGY

CHAPTER VI

Autotrophic Nutrition

1. The types of nutrition found among organisms. Nutrition is that physiological activity whereby organisms make or obtain food. The food provides the *source of energy* for all other metabolic activities and a part of its actual chemical content is necessary for *growth*, *maintenance* and *reproduction* in the organism.

A classification of nutritional methods may be made as follows:

(*a*) *Autotrophic* feeding which applies to organisms that manufacture their food from simple inorganic substances. There are two kinds of autotrophs:

(*i*) *Chemosynthetic autotrophs* use the energy from oxidation of substances already present on earth to "fix" carbon to organic compounds, e.g. sulphur bacteria.

(*ii*) *Photosynthetic autotrophs* use the energy of light to fix carbon to organic compounds, e.g. green plants.

(*b*) *Heterotrophic* feeding applies to those organisms which depend on an organic source of carbon, and which are thus dependent on autotrophs. There are three kinds of heterotrophs:

(*i*) *Holozoic heterotrophs* are those which, like animals, feed on solid material which they take in through a mouth.

(*ii*) *Saprophytic heterotrophs* include organisms such as the Fungi which takes in dissolved organic material, often all over the body.

(*iii*) *Parasitic heterotrophs* ("parasitic" means also a way of life as well as a method of nutrition) are those which feed in or on another organism, the host, and cause it harm, e.g. tapeworm, potato blight.

In this chapter we shall be considering the nutrition of the autotropes in more detail.

CHEMOSYNTHETIC AUTOTROPHS

2. The chemosynthetic autotrophs. As defined above these organisms use the enrgy from the oxidation of substances such as iron or sulphur naturally occurring on the earth's surface. The energy is made available by the same sort of reactions that take place in photosynthesis, that is, the *fixation of the carbon of CO_2 into an organic form*.

These oxidation reactions yield far fewer joules than can be obtained from the utilisation of solar radiation, and although chemosynthetic autotrophs may have been the first type of living organisms they are now much less common. It should however be noted that they play essential roles in the cycling of nutrients in the soil and elsewhere where decomposition occurs.

Some examples are as follows:

(a) *Sulphur bacteria.* There are two types of these bacteria:

(i) *Thiobacillus thio-oxidans* which uses the oxidation of sulphur and water to give sulphuric acid.

(ii) *T. denitrificans* which oxidises thiosulphate with potassium nitrate to give potassium sulphate and release nitrogen. This bacterium, because of its ability to denitrify the soil, leads, under certain conditions, to serious loss of soil fertility.

(b) *Nitrogen bacteria.* These fall into two categories:

(i) *Nitrosomonas* oxidises ammonia, or its compounds, to nitrate. It is an essential link in the nitrogen cycle which leads to the regeneration in the soil of nitrates suitable for plant-root uptake.

(ii) *Nitrobacter* which oxidises the nitrite from the above to nitrate has similar importance.

(c) *Iron bacteria.* These oxidise the iron from the divalent Fe^{++} to the trivalent Fe^{+++} state.

It is clear that these forms of life have been largely superseded by more efficient forms of nutrition. The raw materials upon which they depend are far less available than the sun's radiation and the energy yields of these reactions are also rather small.

PHOTOSYNTHETIC AUTOTROPHS

3. The green plant. The wide variety of terrestrial and aquatic green plants are the most obvious forms of photosynthetic autotrophs although they are not the only forms, for certain bacteria may carry out photosynthesis.

NOTE: "Green" is used here in the sense of "non-white". It therefore includes other colours of photosynthesising plants such as brown algae, etc.

Whether found on land or in the water, most photosynthetic plants tend to have certain features related to their means of nutrition. These are as follows:

(a) A means of *anchorage* or *rooting*.

(b) A means of *uptake for water, CO_2 and mineral salts* and for their *free transport* to the photosynthesising tissues.

(c) Specialised regions with *large surface areas* spread out in the light where the cells are full of plastids with the particular photosynthetic pigment of the plant. The regions have some means of support.

Such features are common from the green algae to the advanced flowering plants, although, as colonisation of the land occurred, the adaptations of plants for efficient photosynthesis, together with adaptations for coping with extensive water losses involved in having large surface areas, grew more complex.

Adaptation to this means of nutrition is seen most highly developed in the *mesophytic leaf* of the flowering plant. Here the leaf is orientated at right angles to the light, a cuticle and stomata regulate water loss and CO_2 entry, and extensive vascular systems bring inorganic ions and water, and remove the final products. Besides this the arrangement of tissues within the leaf allows efficient penetration of light and circulation of gases, and within such tissues the structure of the actual chloroplast is seen at its most refined.

It is worth remembering that the whole somatic structure and growth pattern of plants such as trees is related to their ability to photosynthesise at the maximum efficiency while still overcoming the problem of desiccation imposed by land life.

4. Water uptake by the plant. Water enters the plant through all the unsuberised or unlignified cells of the root. The main means of uptake is by osmosis.

5. The plant cell and osmosis. Osmosis is the movement of water from a high partial pressure to a low partial pressure across a semi-permeable membrane.

(a) *General features.* The cells of the root in contact with the soil provide the conditions for osmosis to occur. The cell sap has osmotically active substances which are separated from the external solution in the soil by the plasma membrane, as well as by the tonoplast, both of which are semi-permeable.

Water is drawn in from the soil and across these two membranes thus entering the cell sap which consequently increases in volume. This increase is finally checked by the tension developed in the cellulose wall, the *wall pressure* (*W.P.*), which prevents the cell from rupture. The force causing the entry of water may be termed the suction pressure, although because of the confusion of ideas in using such a term, *diffusion pressure deficit* (*D.P.D.*) or *water diffusion potential* (*W.D.P.*) are to be preferred.

At any one time the W.D.P., that is, the tendency of water to enter the cell, is expressed by the difference between the actual *osmotic pressure* (*O.P.*) of the cell sap drawing it in and the wall pressure resisting its entry. Thus there is the formula

$$W.D.P. = O.P. - W.P.$$

It can be seen that where the wall pressure and osmotic pressure are equal, the $W.D.P. = 0$. If on the other hand the cell is placed in a solution with a greater osmotic pressure than its own, water will be lost from the cell sap and eventually the protoplast will shrink away from the cell wall. In such cases the walls will resist shrinkage from their normal shapes and a negative wall pressure results. For such situations

$$W.D.P. = O.P. - (-W.P.)$$

i.e.

$$W.D.P. = O.P. + W.P.$$

(*b*) *Turgor and plasmolysis.* When the cell has drawn in water to its full capacity it is said to be fully *turgid*. When water is just being drawn out of the cell it is described as on the point of *incipient plasmolysis* and when the protolast has shrunk away from the cell, it is said to be *fully plasmolysed*. In their normal condition, all the living cells of the green plant are turgid and plasmolysis is only to be found when wilting has taken place. The general relationships of the plant cell to water are shown in Fig. 51.

6. Water entry into the root. This is dependent on surface area and osmosis.

(*a*) *Surface area.* Roots have a very large surface area in contact with the soil and investigations of the roots of cereals indi-

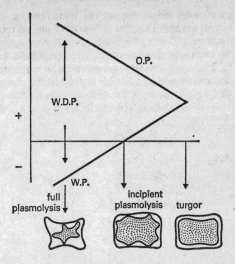

FIG. 51   *The water relation of the plant cell.*

The appearance of the cell at full turgor, incipient plasmolysis and full
plasmolysis is shown. W.D.P. = water diffusion pressure; O.P. =
osmotic pressure; W.P. = wall pressure.

cate that a single plant may have a system several hundred kilometres in length with a new growth measured in kilometres per
day. The surface area of the permeable-cell region is increased
some ten times by the root hairs and they, and the other permeable, cells have walls which provide an extensive network in
molecular contact with soil water.

Plant roots grow into new regions of soil each day to tap its
water and ions and the demand for the former may be very high.
A 10 m mesophyte tree has been recorded to lose up to two
hundred litres of water by a single day's transpiration.

(b) *Osmosis.* The root hairs and other permeable cells have
sucrose and probably other osmotically active substances transported into their cell sap from the upper parts of the plant. These
generate an osmotic pressure which tends to be between five and
six times higher than the capillary forces holding water in the
soil (50–100 kPa). In exceptional cases osmotic pressures in
excess of 10 MPa have been measured for desert xerophytes.

7. Passage of water across the cortex. Water passes across the cortex via two routes. The first is via the fully permeable cell walls and spaces and is a non-living route termed *apoplastic*. It involves no metabolic process and is due to diffusion and capillary forces. The second route is *symplastic* whereby water moves from the cytoplasm and vacuole of one cell to the one next to it (*see* Fig. 52). This depends on the differences in suction pressures between the cells and these tend to increase towards the endodermis.

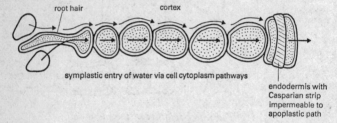

apoplastic entry of water via cell walls and spaces

root hair   cortex

symplastic entry of water via cell cytoplasm pathways

endodermis with Casparian strip impermeable to apoplastic path

FIG. 52 *Apoplastic and symplastic passage of water across cortex of root.*

8. Entry by active uptake. There is a good deal of evidence to suggest that water can also enter the surface cells of the root by active uptake, analogous to that of the entry of certain minerals. Under normal conditions, however, osmotic uptake is the major means of its entry.

9. Passage of water from the root to the leaf. From the soil to the topmost leaf of a tall tree there is a continuous column of water. When the stomata are open water escapes to the air and the whole column moves.

This movement of water within the plant and its escape into the atmosphere is called *transpiration*. In order to understand how the essential nutrient water is brought to those parts of the plant where it can be used in photosynthesis, it is necessary to understand transpiration.

10. Root pressure. In 7–8 it was suggested that there was an active process involved in the transfer of water from the endodermis to the xylem vessels. Such an active process is more noticeable at certain times of the year and it is the pushing of the water upwards by the root that is termed the *root pressure*.

Root pressure may be shown by fixing a glass tube over the excised stem of an actively growing plant. Sap rises in the tube but only for a short distance. Although root pressure certainly exists it is not normally sufficient to account for movement of water up a plant that is more than 10 m or so in height.

TRANSPIRATION

11. Passage from leaf to air. The cells of the leaf are turgid and the cellulose of their walls is also full of water. As these cells are in contact with the spaces in the spongy mesophyll the air in these spaces will also be saturated with water, i.e. will have an *R.H.* near one hundred per cent (*R.H.* = relative humidity).

The atmospheric air outside the leaf may have a variable *R.H.* and this will usually be a good deal lower than one hundred per cent. It is possible to equate *R.H.* (water-vapour content) to values of water diffusion potential thus:

$$R.H.\ 80 = \quad 30 \text{ MPa } W.D.P.$$
$$R.H.\ 55 = \quad 80 \text{ MPa } W.D.P.$$
$$R.H.\ 20 = 200 \text{ MPa } W.D.P.$$

As the *W.D.P.* inside the leaf cells is much less than 10 MPa, even for upper leaves in tall trees, it is quite clear that the opening of the stomata will allow the escape of water to the air and that this "escaping tendency" will often be very high.

12. Factors affecting water loss from leaves. The leaf spaces are fully saturated with water vapour and it is the difference between this concentration and that of the surrounding atmosphere at any given time that provides the driving force in rates of water loss.

In order to escape to the air surrounding the plant the water must pass through two diffusion pathways, each with its own resistance and with its own controlling factors. The final rate of water loss, and thus transpiration, is therefore dependent on many interconnected factors.

(*a*) *The stomatal path* has a resistance to diffusion that depends on the fixed-length characteristic of any given species but varies according to the diameter of the cross section. This in turn depends on the degree of turgor of the guard cells which is usually determined by the light, but also may relate to complex internal rhythms of the plant. Generally, however, where the light is limiting it also is the key factor in determining the rate of water loss from the leaf. Surprisingly enough, once the stomata are open by only a small per cent they no longer influence the rate of transpiration.

(b) *The external path* from the surface of the leaf to the air of normal composition, some way from the influence of the plant microclimate, can be of very variable length.

If the air is still there will be a thick shell of saturated vapour hanging close to the plant surface and diffusion across this will naturally be slow. On the other hand in moving air the layer of saturated water vapour will be very thin and molecules will quickly pass across to the exterior.

For this reason the final rate of water loss from the leaf to the atmosphere will vary considerably in wind and in calm air.

(c) *Factors that interact to determine the rate of water loss* can be expressed by the driving force, essentially represented by differences in concentration of water inside and outside the plant, and the sum of the various resistances to diffusion. Thus

R (rate of loss)

$$\propto \frac{\text{Water conc. inside leaf} - \text{Water conc. in air}}{\text{Stomatal resistance} + \text{Water shell resistance immediate to}}$$
$$\text{leaf surface}$$

The way in which the interaction of the various factors works out in practice is seen in Fig. 53, which relates to water loss under different environmental situations.

13. The cohesion theory. Water movement through the xylem is through the capillary tubes of the xylem which are very thin, some 0·4 mm on average, and it can be demonstrated that pressures in excess of 30 MPa are required to break such thin columns. This is provided that no bubbles of air are present. It is suggested that a complete and unbreakable column of water is drawn up the plant through the tracheids and younger vessels. The force for this is supplied by the escaping tendency or evaporation of the water from the leaves.

NOTE: While some experimental evidence supports such a view it is by no means conclusive and living parenchyma in the xylem may well have an important role in maintaining water within the conducting elements and preventing formation of

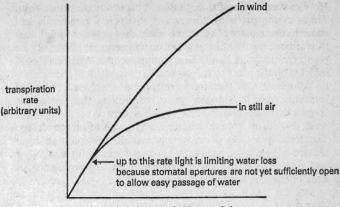

FIG. 53   *Water loss in plants under different environmental situations.*

air bubbles. The larger vessels are almost certainly full of air and thus play no part in conduction.

14. Resistances to water movements. In the first place there is the gaseous path out of the leaf which offers resistance to water movement. This path represents a resistance some twenty times that of all the others put together.

NOTE: The xylem path depends on the number and size of active elements. Rates of flow may exceed 200 cm per hr/cm³ in mesophytes whereas in xerophytes they may be rather less than 1 cm per hour.

The path from the root to the soil will have a resistance dependent on the relative suction of the water-absorbing cells and the availability of water in the soil.

If the resistances, all together exceed the escaping tendency, no transpiration will take place. If the loss from the leaves is too great to be met by flow into the leaves or up the xylem, wilting will occur. It, as is rather unlikely, water intake by the roots exceeds water loss from the leaves then *guttation*, or dripping of water out of the leaves will occur. This is not uncommon in humid tropical climates.

15. The significance of transpiration. This is not altogether clear. Transpiration provides *a means of transport* for minerals and water to those parts of the plant where they are used and it does this without the plant having to expend energy. It also cools the plant. On the other hand transpiration seems a "necessary evil" inflicted on the plant by its physical nature. Much of the successful colonisation of the land by plants has involved a means of satisfying and, to a lesser extent, controlling this process. It should be noted that a plant uses less than 1 per cent of the water it takes up from the soil for the synthesis of carbohydrates. The total quantities of water involved in transpiration by a hectare of vegetation can come out at millions of litres in a year and there is no doubt that the transpiration of green plants has a major effect on the whole climate of a region.

ᐧUPTAKE OF MINERALS

16. The uptake of mineral ions by the plant. The same surfaces that operate for the uptake of water also serve for the entry of ions into the plant. These ions are partly derived from the soil water and partly from exchanges with the colloidal micelles of clay particles on which cations tend to be held. Ultimately the *main source* of nutrient salts is from the breakdown of protoplasmic substances by the activity of saprophytes in the soil.

At one time it was thought that many ions diffused into the permeable cells of the root along their concentration gradients which were maintained by the continual removal of the ions up the plant. It is now clear that the majority of ions do not enter by simple diffusion but that *active or exchange mechanisms* are involved.

17. Pathways into the root. In the first place ions are thought to move from the soil across the cellulose walls of root cells. The plasma membrane offers little resistance to their movement and they pass into the cytoplasm of the cells.

(*a*) *Transport across the tonoplast.* This is an active process involving expenditure of metabolic energy and by this means ions become concentrated within the cell vacuole. The concentrations may reach as much as one thousand times that of the ion in the external medium.

Movement from cell to cell across the cortex also involves

active pathways and finally the ions are pumped into the xylem elements whence they can be transported up the plant.

(b) *Mechanisms of active transport.* It is not possible to be sure exactly how active transport of ions takes place but it has been suggested that the ion required is joined at the membrane surface to a carrier molecule. This circulates to the inside of the membrane where the ion is released into the cytoplasm or

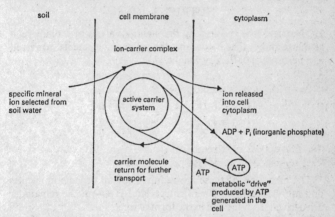

FIG. 54   *Hypothetical model of active transport across root cells.*

vacuole, whichever the case may be. The energy involved for the loading and circulation of the carrier is made available from ATP from aerobic oxidation. Roots require a substantial supply of oxygen to maintain ion uptake.

At the same time as a cation, such as Na^+, Ca^{++} or K^+, is taken into the plant, anions are also taken up. These will be OH' or HCO_3' and by this means the balance of charges within the plant is maintained. Conversely uptake of the anions H_2PO_4', NO_3', SO_4'', etc. is accompanied by uptake of cation H^+.

Another means of ion uptake suggested is that of exchange whereby the plant exchanges an anion or cation that is not required for one that is. Ions moved out of the plant in this way might include H^+, OH', HCO_3', or Na^+.

18. Movement of ions within the plant. Some ions will combine directly with organic substances in the root cells and be transported to other parts of the plant as complex molecules. The majority remain in their original form and are carried up the plant in the transpiration stream, while there is also a much smaller movement of ions in the phloem (see Table VII). There is evidence for the translocation of some ions from the xylem to the phloem during transport.

NUTRIENT IONS

19. Nutrient ions required by the plant. Analysis of a plant such as maize shows that it contains the following elements expressed as a percentage of the dry weight.

C	43·5	Ca	0·2
O	44·4	K	0·9
H	6·2	Mg	0·2
N	1·5	Fe	0·1
S	0·2	Mn	0·03
P	0·2	Si	1·2
Al	0·1	Cl	0·1

The first three of these are required for all organic molecules and are variously obtained as CO_2 and H_2O but the remainder are derived from the soil in the form of ions.

These elements are used by the plant for the *synthesis* of certain protoplasmic substances, as *co-factors for enzymes, structurally* for cell walls, and for *maintenance of an ionic medium* in which protoplasm can continue to function.

Those that are needed in fairly large quantities are called major elements and those where very small amounts are needed are termed minor elements. Where a particular element is lacking the plant develops a specific mineral-deficiency disease which may be recognised by its symptoms.

20. Uses and deficiencies. The uses and deficiencies caused by the absence of these elements is set out in Table VII.

Because of the marked effect that the above nutrients have on the growth of plants it is customary to ensure an optimum supply to crops by the application of fertilisers. A selection of these contain all the major elements required. Thus N, S, Na, K are supplied in ammonium sulphate and sodium or potassium nitrate, P, Ca are contained in superphosphate and calcium phosphate and Mg is found in magnesium phosphate.

TABLE VII.   USES AND DEFICIENCIES OF NUTRIENT IONS IN THE
PLANT

Element	Form in which taken up by the plant	Use in the plant	Nature of deficiency
N	NO_3'	Incorporated into amino-acids.	Poor growth. Chlorosis.
S	SO_4''	Incorporated into amino-acid cysteine, cystine and methionine. Also used as cross bridges in the tertiary structure of many proteins.	Excessive root growth but poor growth in other parts. Chlorosis.
P	H_2PO_4'	Used in membrane synthesis, nucleotides and energy carriers.	Poor growth. Failure of transport and respiratory systems.
Ca	Ca^{++}	Used by some plants in middle lamella of cell walls. In all plants, important for normal membrane functioning.	Failure of turgidity and thus poor growth.
Fe	Fe^{++}	Precursor in chlorophyll synthesis, involved in respiratory enzymes, e.g. cytochrome.	Chlorosis. Reduced growth.
Mg	Mg^{++}	Incorporated into the chlorophyll molecule. A co-factor in the phosphorylating enzymes. ATP action.	Chlorosis. Reduced growth.
K	K^+	Necessary for activity of certain enzymes and for membrane functioning.	Reduction in photo-synthesis, a premature yellowing of the leaves.

Minor elements include Zn, Mo, B, Mn, Cu, Cl, etc. Many of these are needed in minute amounts as co-factors for specific enzymes. Their lack produces various symptoms but in most cases the growth of the plant is much reduced.

CARBON DIOXIDE

21. The entry of carbon dioxide into the plant. A small amount of carbon enters the plant in the form of HCO_3' derived from the soil. The major source is from the *atmosphere* which contains some 0·04 per cent of the gas carbon dioxide.

The CO_2 is taken in via the stomata from which it is diffused across the air spaces of the leaf into solution at the surface of the photosynthetic cells. The dissolved CO_2 must then pass through the protoplasm of these cells until it reaches the chloroplasts where actual carbon fixation occurs.

22. Structure and function of stomata. The stomata of dicotyledons are surrounded by two guard cells which are differentially thickened and, unlike other cells of the epidermis, have protoplasts that contain chloroplasts. Enlargement of the guard cells (as described below) takes place asymmetrically owing to this thickening and the cells draw apart. The gap between the guard cells is the stomatal pore (*see* Fig 55).

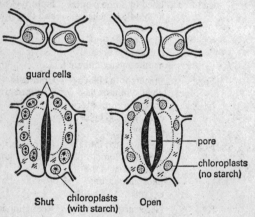

FIG. 55 *Changes in the shape of the guard cells bring about the opening and closing of the stoma.*

NOTE: Although the increasing turgor of the guard cells is the immediate reason for the opening of the stomata, the method by which this turgor is brought about is by no means certain. A possible mechanism is outlined below:

Light: → general initiation of photosynthesis in the leaf:

↓

reduction of CO_2 from the leaf spaces;

↓

loss of CO_2 from the guard cells leading to an increase in the pH of these cells;

↓

enzymic conversion of guard-cell starch into hexose sugars which takes place in alkaline conditions;

↓

presence of sugar in the guard cells increases their osmotic pressure so that water is drawn in from other epidermal cells;

↓

the guard cells become turgid and the stomata open.

The closure of the stomata on such a basis would be due to an increase in the leaf space of CO_2, such as would occur when respiration was in excess of photosynthesis in darkness.

If even the most typical plant (in the sense of opening its stomata in light and closing them in the dark) is studied the changes and determining factors are found to be complex. Thus for such a "typical" plant the relationship between stomatal aperture, light and dark, and amount of starch in the guard cells turns out to be as shown in Fig. 56.

Recent investigations indicate that stomatal opening and

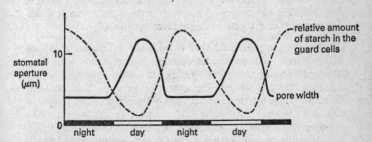

FIG. 56 *Relationship between stomatal aperture, light and dark, and amount of starch in guard cells of plant stomata.*

closing can be influenced by a number of internal and external factors, as shown in Table VIII.

TABLE VIII.   FACTORS AFFECTING STOMATAL OPENING AND CLOSING

Factor	State of factor	Effect on stomata
Relative humidity	Low	Closure
	High	Opening
Concentration of CO_2	Low	Opening
	High	Closure
Temperature	Low	Closure
	Medium	Opening
	High	Closure
Time from previous opening (even in dark)	Short	Tendency to stay shut
	Long	Tendency to open
Time from previous closing (even in dark)	Short	Tendency to stay shut
	Long	Tendency to open

In conclusion it can be seen that the actual stomatal rhythms of a given plant are determined by the complex interaction of a number of internal and external factors and it is not always easy to see the precise survival value of a particular response. Doubtless selection has conferred survival in these responses and the whole area is one of active current investigation.

23. Numbers of stomata. This is normally estimated at about 400 per mm^2 (e.g. apple). Xerophytes, in connection with their ability to reduce transpiration may have only between ten and fifteen stomata per mm^2. The spaces between stomata are such as to minimise interference between diffusion shells of the gases that pass in and out of the leaf. Because of this very precise spacing the numerous tiny pores act much more efficiently in allowing the passage of gases than a smaller number of much larger pores might have done.

THE PROCESS OF PHOTOSYNTHESIS

24. Introduction to photosynthesis. Although simply represented in elementary textbooks the processes of photosynthesis are very complex. At least two stages are involved and these are called the *light* and the *dark reactions*; evidence for their existence comes from a number of studies.

(*a*) *Flashing light.* It was noticed that photosynthesis continued for a short time after a flash of light, also that the amount of carbon fixed was greater for a plant exposed to intermittent light than for one exposed to continuous light for the same period.

This supported the idea that a "capacity" reaction was being saturated by the light and a slower reaction was proceeding in the dark utilising the products of the light reaction.

(*b*) *Temperature.* The effects of temperature were investigated when either light or carbon dioxide were limited. For the former, with optimum CO_2, temperature changes did not affect the rate of the reaction, while with limited CO_2 the effects of temperature were very marked.

These results can be interpreted by supposing that the first reaction is *photochemical*, as one of the characteristics of such reactions are that they are unaffected by temperature. The second, or dark reaction, which involved CO_2, was a normal *thermochemical* reaction.

(*c*) *Isotopes.* Further evidence came from the use of a heavy isotope of oxygen ^{18}O incorporated into water as $H_2{}^{18}O$. When this was used the oxygen released by the plant during photosynthesis was ^{18}O showing that it must have come from the water.

The really important breakthrough in understanding the chemical pathways of photosynthesis came from the work of Calvin in America who used ^{14}C to follow carbon fixation.

25. The light reaction. This is the first stage of photosynthesis and is a photochemical reaction whereby light energy is captured by the pigment systems in the chloroplast and converted to chemical energy.

(*a*) *The roles of the different pigments.* While chlorophyll *a* is the most important of the photosynthetic pigments others with different absorption spectra make use of other parts of the spectrum and transfer the energy they obtain to the chlorophyll *a*. In this way green plants are able to utilise quite a wide range of

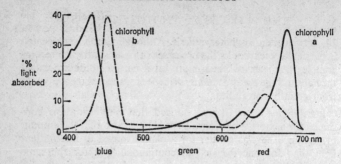

FIG. 57 *Absorption spectra of chlorophyll* a *and chlorophyll* b.

light wavelengths. The absorption spectra of the two chlorophylls is shown in Fig. 57.

Another important pigment is carotene but it now appears that the main importance of this is to take up excessive quantities of light at high intensities and thus prevent destruction of the vital chlorophylls.

As we have seen in V, **16,** the chlorophyll molecules are arranged in layers in the thylakoids of the grana of the chloroplasts. It seems that groups of individual molecules work together to catch quanta of light energy, Thus, some 300 molecules trap several quanta initially and these are "concentrated" and passed on to a system involving about ten chlorophyll *a* molecules. Finally the energy, now in a very concentrated form, passes to a single chlorophyll molecule which becomes activated and from which the electron transfer starts (*see* below).

(*b*) *Cyclic phosphorylation.* Light of less than 690 nm is taken up by chlorophyll *b* which becomes photo-excited. Its energy level is raised so that a molecule of water is split into its components, the oxygen being released and the hydrogen reducing the chlorophyll *b*. The chlorophyll is activated during this process and its reducing potential raised above that of chlorophyll *a*.

The electrons carried by the chlorophyll *b* are now transferred through an electron transfer system involving a substance called plastoquinone and cytochromes *b* and *f* (this latter being unique to this process).

Two results occur:

(*i*) During the electron transfer process a low energy phosphate is activated and transferred to an ADP, molecule thus generating a molecule of high energy ATP. (We shall see that this ATP is essential for the carrying out of the dark process of carbon fixation a few hundreds of a second later.)

(*ii*) The chlorophyll *a* is itself activated partly by energy transfer from the electron chain and partly by direct photo-excitation by light of the 690 nm wavelengths and above.

This whole sequence is called cyclic phosphorylation because at the end of it an ATP has been generated and the electrons returned to the original chlorophyll *b* molecule where the whole process can recommence. The process thus far is indicated in Fig. 58.

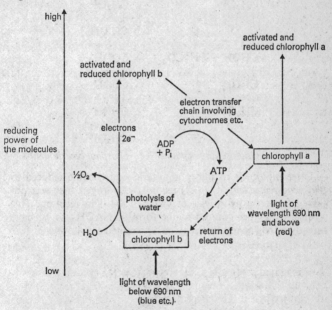

FIG. 58 *Diagram of cyclic phosphorylation.*

(*c*) *Non-cyclic phosphorylation.* This is still part of the light reaction and it involves the formation of the powerful hydrogen transfer molecule NADPH₂ (nicotinamide adenine dinucleotide phosphate—reduced state. *See also* VIII, 5).

The activated and reduced chlorophyll *a* formed, as already explained above, passes on its excited electrons (which join up with the protons from the original water molecule hydrogen to form hydrogen) from ferredoxin to flavoprotein and thence to NADP, giving $NADPH_2$.

From this key substance the hydrogen can enter an intermediate organic molecule in the first part of the dark reaction of photosynthesis.

26. The dark reaction. The CO_2 as the soluble carbonic acid in the chloroplasts becomes incorporated with a 5-carbon compound called *ribulose diphosphate* (*RDP*) to give two molecules of *phosphoglyceric acid*:

$$
\begin{array}{ll}
\begin{array}{l}
CH_2OP \\
| \\
C{=}O \\
| \\
CHOH \\
| \\
CHOH \\
| \\
CH_2OP
\end{array}
+ H_2CO_3 \rightarrow
&
\begin{array}{l}
CH_2OP \\
| \\
CH.OH \\
| \\
COOH \\
COOH \\
| \\
CHOH \\
| \\
CH_2OP
\end{array}
\end{array}
$$

ribulose diphosphate   *phosphoglyceric acid*

These molecules of phosphoglyceric acid become reduced to *phosphoglyceraldehyde* (*PGA*) using the hydrogen that was obtained from the hydrolysis of water in the light reaction. This hydrogen has been stored in the form of $NADH_2$ as we have already seen. The reaction may be represented as follows:

$$
\begin{array}{l}
CH_2OP \\
| \\
CHOH \\
| \\
COOH
\end{array}
+ NADH_2 \rightarrow
\begin{array}{l}
CH_2OP \\
| \\
CHOH \\
| \\
CHO
\end{array}
+ NAD + H_2O
$$

phosphoglyceric acid   *phosphoglyceraldehyde*

Phosphoglyceraldehyde is a key 3-carbon substance which is the basis of other syntheses which themselves should properly be regarded as a part of the photosynthetic process.

Some of the fates of this substance are set out in Fig. 59 and the more important of these are considered in more detail in the sections that follow.

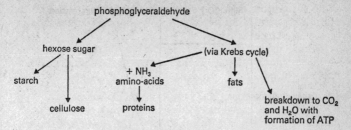

FIG. 59 *Phosphoglyceraldehyde as the basis for other photo-synthetic processes.*

All these reactions take place within seconds of the incorporation of the original carbon and all are part of photosynthesis.

In order for the carbon fixation to continue a supply of the acceptor substance ribulose diphosphate is necessary and this is also derived from a large part of the phosphoglyceraldehyde formed through two intermediates which have four and seven carbon atoms respectively. The process of regeneration of ribulose is called the Calvin cycle after its discoverer.

27. Regeneration of ribulose. Ribulose can be regenerated by a whole variety of different routes involving all sorts of small CHO intermediates, as shown in Fig. 60. Perhaps having so

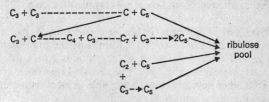

FIG. 60 *Formation of ribulose pool.*

many paths, each with its own particular enzymes, ensures that the supply of the essential ribulose, the only acceptor of CO_2, never gets depleted.

The interrelationship between phosphoglyceraldehyde and ribulose diphosphate is as shown in Fig. 61. Where CO_2 is not limiting the dark reaction the concentration of PGA is high as it is being continuously made from RDP, whose own concentra-

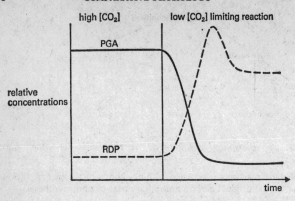

FIG. 61   *Relationship between phosphoglyceraldehyde (PGA) and
ribulose diphosphate (RDP).*

tion is therefore low. When CO_2 is very low and limiting the
dark reaction, then little PGA can be synthesised and the RDP
is not being used: its concentration is therefore high. Thus, concentrations of RDP and PGA tend to be inversely proportional
and depend on the internal environmental factors operating at
any given time.

28. Schematic representation of photosynthesis. (See p. 127)

29. Synthesis of nitrogen-containing compounds. A part of the
phosphoglyceraldehyde or PGA formed as the immediate product of the dark reaction of photosynthesis enters the Krebs
cycle. At one stage in this cycle it becomes *α-ketoglutaric acid* and
this is another key substance, being able to react with ammonia
to form the amino-acid *glutamic acid*:

$$
\begin{array}{l}
\text{COOH} \\
\mid \\
\text{(CH}_2\text{)}_2 \\
\mid \\
\text{CO} + \text{NH}_3 + \text{NADH}_2 \\
\mid \\
\text{COOH}
\end{array}
\rightarrow
\begin{array}{l}
\text{COOH} \\
\mid \\
\text{(CH}_2\text{)}_2 \\
\mid \\
\text{CHNH}_2 + \text{NAD} + \text{H}_2\text{O} \\
\mid \\
\text{COOH}
\end{array}
$$

α-ketoglutaric acid   glutamic acid

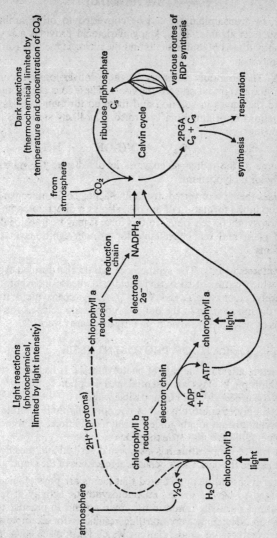

FIG. 62 Schematic representation of photosynthesis.

This, by transamination, can be converted to other amino-acids such as alanine. Thus, ketoglutaric acid provides a route into the synthesis of amino-acids and the nitrogen metabolism of the plant.

NOTE: The ammonia necessary for these conversions has come from the NO_3' that entered the roots. These ions are brought up to the leaves in solution and there go through a series of reductions until ammonia is formed. The likely steps are the following:

$$NO_3' \rightarrow NO_2' \rightarrow NH_2OH \rightarrow NH_3$$

Some of the reductase enzymes involved need the mineral co-factor molybdenum.

Besides being converted to amino-acids some nitrogen is taken up in the formation of amides, also as a direct outcome of the photosynthetic reactions. In this form it may be translocated in the plant and made available for protein synthesis in the meristems.

30. Synthesis of fats. The synthesis of fats is also formed from PGA which is turned to pyruvic acid and thence incorporated into acetyl co-enzyme A (*see* VIII, 7). Acetyl coenzyme A may act as the initial stage in the building up of fatty acids as the reactions by which it is formed from fatty acids are reversible.

RATE OF PHOTOSYNTHESIS

31. Factors affecting the rate of photosynthesis. It has been seen that chlorophyll plays an essential part in photosynthesis and clearly the process would be impossible without it.

The supply of chlorophyll does not normally limit the rate of photosynthesis although under certain conditions of mineral deficiency chlorosis may stop the process.

The role of temperature has also been described and the effects that it has on the dark thermochemical reactions of the plant.

NOTE: It should be appreciated that plants are physiologically adapted to have a maximum photosynthetic activity at temperatures normally found in the environment. In practice this may produce some very startling results as for example the photosynthesis of arctic diatoms which takes place efficiently at temperatures below 0°C.

Other important factors are light and carbon dioxide.

32. Light. If the carbon dioxide level is the optimum, the effects of varying the light intensity on the rate of photosynthesis can be measured. It is found that the rate is directly proportional to the light intensity up to the point of saturation. Some plants naturally live in shade and these reach their maximum rates of photosynthesis at relatively low-light intensities. Others live exposed to full sunlight and only reach maximum levels at high intensities.

The compensation point of a plant is where the sugar gain from photosynthesis is exactly equal to the loss by respiration. In shade plants it is very much lower than in sun plants but the former are not able to reach the high levels of photosynthesis achieved by the latter.

In fact in a large green plant such as a bush or tree the outer leaves (called sun leaves) have a different compensation point to the inner leaves (called shade leaves). The effect of this is to improve the overall efficiency of the whole plant in bright sunlight, as when the outer leaves are saturated with light and cannot increase carbon fixation the inner ones are able to take up the excess light coming through and make use of it with their much lower demands. Thus a whole tree will seldom if ever be fully saturated with light but can adapts itself from moment to moment to make the best use of the incident wavelengths and intensities.

33. Carbon dioxide. Here the light is set at optimum and the effect of changing the CO_2 measured. Once again the rate of photosynthesis is found to be proportional to the $[CO_2]$ until approximately one per cent where it begins to exert a depressive effect on the plant's respiration.

It is interesting to note that the concentration of CO_2 in the atmosphere is only about 0·036 per cent, which is far below the optimum value for photosynthesis.

34. Experimental evidence for the photosynthetic reactions. The rates of photosynthesis in Fig. 63 are obtained under different conditions of light intensity, temperature and carbon dioxide concentrations. If they are analysed in depth they support the concept that photosynthesis is a two-stage reaction. Thus:

(a) At low light intensities the rate of photosynthesis is not affected by temperature. This is a characteristic of photochemical

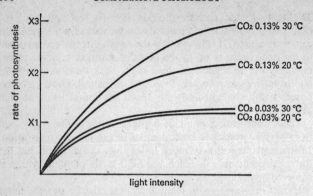

FIG. 63 *Effect of light intensity, concentration of* CO_2 *and temperature on the rate of photosynthesis.*

reactions involving light. At this stage the light reaction is limiting the rate of photosynthesis.

(b) At high light intensities, that is to the right of the graph, the dark reaction involving CO_2 is the limiting factor. This is a thermochemical reaction and is temperature dependent. Thus where CO_2 levels are high at 0·13 per cent the reaction takes place much more rapidly at 30°C than at 20°C. (Normally biollogical reactions show a doubling for a 10°C increase in temperature.)

Where CO_2 concentration is itself limiting the process, as in the two lower curves, the reaction is not affected by change in temperature.

35. Measuring the rate of photosynthesis. There are several possible methods for doing this:

(a) *Changes in the dry weight.* If portions of a leaf are removed with a cork borer and heated to constant weight and then similar portions taken after the plant has been photosynthesising, the apparent gain in weight is partly due to photosynthesis. Respiratory losses over the same period (*see* 28) have to be measured and added to the apparent gain and there is some loss due to translocation (*see* 37).

(b) *Carbon dioxide uptake.* This may be measured gasometrically with a manometer or by passing a known concentration of CO_2 over the plant and estimating the quantity taken up.

(c) *Production of* O_2. This may be measured in a photosyntho-meter (*see* Fig. 64). This method is most suitable for aquatic plants such as *Elodea* and the apparatus allows the quantitative measurement of the O_2 released in a given time.

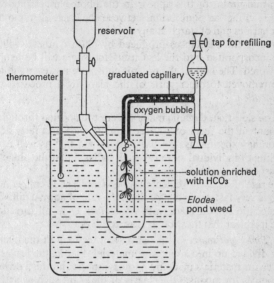

FIG. 64. *The photosynthometer is used to measure the rate of photosynthesis by exact measurement of the oxygen produced.*

The temperature and light intensity can be regulated at will.

In both (b) and (c) suitable corrections must be made for gas exchanges due to respiration if an actual value of photosynthesis is to be obtained.

(d) *Use of* $^{14}CO_2$. This is yet another method of estimating the rate of photosynthesis and it has been used extensively in studies of phytoplankton and the productivity of the seas.

FATE OF THE PRODUCTS OF PHOTOSYNTHESIS

36. The products of photosynthesis. As described above the car-bon from the atmosphere is fixed with ribulose diphosphate to form two molecules of phosphoglyceraldehyde or PGA. This compound may then be converted to various types of carbo-hydrate as well as fats and with NH_3 it may be converted into amino-acids and thus proteins.

An immediate product in most plants is the polysaccharide starch and grains of this appear in the photosynthesising cells. There is in fact an equilibrium between the levels of mono- and disaccharides and of starch at any one time.

Some of the substances produced by photosynthesis will be-come incorporated into the leaf itself and some will be tempor-arily stored. The bulk of the organic substance however will tend to be removed to other parts of the plant by a process called *translocation*.

37. Translocation. Organic materials are transported throughout the plant in the form of sucrose, amino-acids and fatty acids. Sectioning experiments, the use of dyes and C^{14} as well as the sampling of individual sieve tubes all indicate that the phloem is the tissue responsible for translocation.

The rate of sugar movement may be as much as 50 cm per hour, which is very fast, some 60,000 times the rate of normal diffusion. Translocation takes place with equal facility in all directions, hence it is clear that the process is an active one.

(a) *Sinks and sources*. It is convenient to think of the plant as being divided up into sources of material, that is, the leaves are sources for materials made by photosynthesis and the meristems are sources for auxins. Other parts of the plant such as actively growing regions and storage organs or the fruits and seeds may act as sinks into which the translocated material is moved. Once it has reached its particular sink the soluble organic molecules are utilised in respiration or synthesis or else condensed into a form suitable for storage.

(b) *The mechanism of translocation*. This is by no means under-stood and the most widely held view, that of a pressure-flow system, does not fit in well with what is known of the actual structure of the sieve tubes. As can be seen from Fig. 65, a sieve tube is always associated with a companion cell: the integrity of the latter seems necessary for phloem translocation to occur. This lends some support to the idea of an ion-pumping mechanism in active transport. Interpretation of the structure as shown by electron microphotographs is still controversial.

NOTE: The pressure-flow system suggests that the source areas of the plant tend to have a high osmotic pressure while the

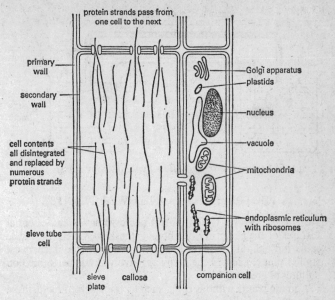

FIG. 65 *A sieve tube and its companion cell.*

sinks have a much lower one. Since the two areas are in direct protoplasmic communication through the sieve tubes of the phloem it is thought that organic substances are forced along the pressure gradient. At the same time water will circulate up the xylem and to a much lesser extent will circulate back through the phloem.

Although this pressure-flow hypothesis is attractive it is by no means compatible with all the experimental data available. A model of this system is shown in Fig. 66. If this is turned on its end it may be considered to represent the actual state of affairs existing within the plant.

Other theories suggest either movement of the cytoplasm rather similar to cyclosis in photosynthetic cells or else the active transport of ions such as K^+ by companion cells producing an electrical gradient along which sugars will move. Again some interpretations of sieve tubes indicate the presence of microtubules. Clearly the means of phloem transport is not yet established.

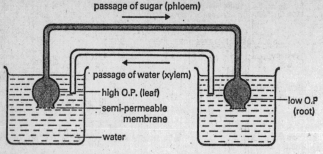

FIG. 66   *Munch's pressure-flow hypothesis for translocation.*

PHOTOSYNTHESIS IN LOWER PLANTS

38. Special features of lower-plant photosynthesis. The essential features of the chemistry of photosynthesis seem very much the same throughout the plant kingdom. Much of the biochemical studies that elucidated the pathways involved was performed on cultures of green algae.

Aquatic plants without roots such as the marine fucoids have special hairs, called cryptostomata, which are associated with mineral-salt uptake.

Again the complexity of organisation found in the mesophyte leaf is not present in plants below the level of pteridophytes. Bryophytes have neither stomata nor cuticle so tend to have considerable problems of water economy.

Algal chloroplasts tend to be very large and contain pyrenoids which are centres for starch condensation.

PROGRESS TEST 6

1. List the major types of nutritional methods used by living organisms.   (1)

2. What is meant by "W.D.P.", "turgor", "incipient plasmolysis"?   (5)

3. What is transpiration?   (9)

4. How does relative humidity affect transpiration rate?   (12)

5. What is the role of the tonoplast in mineral uptake?   (17)

6. What use does the plant make of calcium, iron, magnesium? (19)

7. Of what use were radioactive isotopes in following metabolic pathways of photosynthesis? **(24)**

8. What are the essential stages of the light reaction of photosynthesis? **(25)**

9. What happens in the dark reaction of photosynthesis? **(26)**

10. At what stage may nitrogen be incorporated into organic molecules during photosynthesis? **(29)**

11. How does light affect the rate of photosynthesis? **(32)**

12. What is a photosynthometer and in what units does it measure? **(35)**

13. Outline a hypothesis that could account for the translocation of sugar in plants. **(37)**

EXAMINATION QUESTIONS

The figures in **bold** type indicate the marks allocated to each question or part question.

1. Write brief notes to distinguish between the following parts of a young root:

(a) the apical meristem and the elongating zone, **(3)**

(b) the piliferous layer and the endodermis, **(3)**

(c) the cortex and the stele. **(3)**

Describe concisely the mechanism of water uptake in an intact, transpiring plant. **(9)**

(*Total* **18**) *Cambridge*, 1977

2. By means of diagrams of a leaf and a cell, show where the chloroplasts are situated. Draw a further diagram to show the structure of a chloroplast. Indicate on your diagrams how the raw materials of photosynthesis reach the site where the reactions take place. **(9)**

Describe what is meant by the *light reaction* and the *dark reaction* of photosynthesis. **(9)**

(*Total* **18**) *Cambridge*, 1977

3. Make a large, labelled diagram to show the structure of a phloem sieve tube and a companion cell, as seen with the light microscope. **(4)** List additional structures that could be seen with an electron microscope. **(2)**

What evidence is there that phloem is concerned in solute transport? **(6)** Outline the various hypotheses that can be advanced to explain the mechanism of transport. **(6)**

(*Total* **18**) *Cambridge specimen paper*, 1976

4. The autoradiographs illustrated below were developed

from a suspension of algae which had been incubated in bright light and whose culture had been injected with ^{14}C bicarbonate solution for different times before a sample was run off and fixed immediately in boiling alcohol.

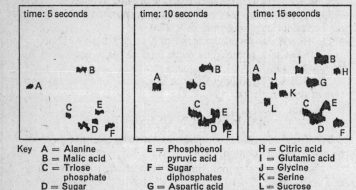

time: 5 seconds   time: 10 seconds   time: 15 seconds

Key			
A = Alanine	E = Phosphoenol	H = Citric acid	
B = Malic acid	pyruvic acid	I = Glutamic acid	
C = Triose	F = Sugar	J = Glycine	
phosphate	diphosphates	K = Serine	
D = Sugar	G = Aspartic acid	L = Sucrose	
phosphates			

Data after Bassham J.A. 1962.

Interpret the results in terms of current ideas on the sequence of reactions involved in carbon fixation in algae during photosynthesis. Your answer should refer only to the products indicated on the autoradiographs.

(Total 20) *Associated Examining Board*, 1974

5. Describe what happens (a) during the light stage and (b) during the dark stage of photosynthesis. (14) Present the evidence that led to the hypothesis that there is a light and dark stage in photosynthesis. (6)

Oxford and Cambridge, 1976

6. Give a brief account of photosynthesis in green plants and show how the concentration of CO_2 may affect the rate of the process of photosynthesis.

Oxford and Cambridge, 1973

7. Describe in detail how water in the soil may reach the xylem of a root. Discuss the possible mechanisms and pathways.

Oxford and Cambridge, 1973

8. Define transpiration. (3) Give an account of the ways in which air and soil conditions affect the rate of transpiration by a plant. (17)

Oxford and Cambridge, 1977

9. List the mineral requirements of a green plant and explain the physiological importance of any *four* of the required ions.

after London, 1975

10. Briefly compare the functions of phloem and xylem in a flowering plant. What kinds of cells are found in the xylem and how is their structure related to the functions of the tissue?

London, 1975

11. The removal of a narrow ring of tissue from the outside of a woody stem leads to the accumulation of organic compounds in the region above the ring. (*a*) Give an explanation for this and name some of the compounds that may accumulate. (*b*) Explain what might happen to these compounds in an intact stem. (*c*) How would water and mineral ions from the roots be affected by the ringing operation? (*d*) How would you demonstrate that (*i*) oxygen and (*ii*) temperature affect the rate of movement of organic compounds in a stem?

London, 1975

12. Write an account of the ways in which external factors may influence the rate of photosynthesis in a green plant. Describe carefully how you would measure (*a*) the rate of photosynthesis and (*b*) the effect of *one* external factor on this rate.

London, 1976

13.

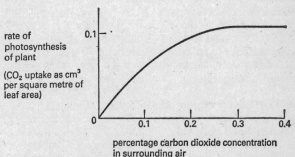

The graph above shows the relationship between the rate of photosynthesis of a flowering plant and concentration of carbon dioxide in the surrounding air.

(*a*) Describe carefully the relationship shown by the graph.
(*b*) What is meant by the concept of "limiting factors"? In

what circumstances could light be a limiting factor for photo-synthesis?

(c) Describe briefly the pathway by which carbon dioxide passes from the air to the site of photosynthesis. Upon what factors does the rate of carbon dioxide uptake depend?

London, 1976

CHAPTER VII

Heterotrophic Nutrition

1. Definitions. Heterotrophic nutrition has been defined in IV, **1**. It will be remembered that heterotrophs are unable to synthesise organic molecules from simple inorganic molecules whereas autotrophs can do this. It will also be remembered that heterotrophs fall into three classes:

(*a*) *Holozoic organisms*, which feed on solid food which they take in through a mouth. They include most animals.

(*b*) *Saprophytic organisms*, which take in soluble substances all over the body surface. They include most fungi and bacteria.

(*c*) *Parasitic organisms*, which live in or on a living host to which they cause some sort of harm.

Each of the above classes will be considered separately in this chapter.

GENERAL FEATURES OF HOLOZOIC NUTRITION

2. The dietary requirements of holozoic organisms. The dietary requirements of animals consist, obviously enough, of the types of chemicals of which their bodies are made (*see* IV for a consideration of these chemicals). The majority of animals require a balanced diet containing *carbohydrates, fats, proteins, vitamins, mineral salts and water*. Besides the quality of diet it is clear that the organisms must have a sufficient quantity to provide the energy for their metabolism.

While the information in IV is relevant to this section there are a few further comments to be made about each item of diet. They are as follows:

(*a*) *Carbohydrates*, usually in the form of starch, provide the main source of carbon for many animals but not for specialised carnivores which use the carbon from protein. Herbivores may take in plant material including cellulose but under the conditions in the rumen these are changed to fatty acids by symbiotic bacteria. Certain flagellate protozoa such as *Polytoma* are able to utilise less complex carbon compounds than carbohydrates, for example acetates, CH_3COOR.

(*b*) *Fats* can be synthesised from carbohydrates or from deaminated proteins in many organisms including certain mammals, insects and protozoa. In other cases a single type of fatty acid is sufficient for the organism to synthesise others that it might require. Other organisms, such as man, are more demanding.

(*c*) *Proteins* are assimilated by animals in the form of amino-acids. There are some twenty of these commonly utilised by animals and those that a particular animal cannot synthesise for itself are essential in its diet.

Certain protozoa such as the colourless forms of *Euglena* can synthesise all their amino-acids from the provision of a single one in their diet. Others such as *Polytoma* show even more remarkable powers of synthesis of nitrogenous compounds and are satisfied with NO_3 or NH_3 as their only source of nitrogen.

Higher organisms are very much more demanding of amino-acids, and such widely separated animals as men, rats and mosquitoes all require approximately half the amino-acids. The above is also applicable to nucleic acids.

(*d*) *Vitamin* requirements are also very variable among holozoic organisms. Vitamins such as A, D, E and K which are essential for many mammals, including man, are not required at all by most invertebrates. Most holozoic organisms require vitamins of the B group.

(*e*) *Mineral salts* that act as enzyme co-factors are described in IV, **20**. Besides these most animals make use of K^+ and Na^+ for membrane pumps and potentials and these and Cl' and other ions for the maintenance of an ionic medium in which enzymes can function. Ca^{++} and PO_4''' are commonly used structurally.

(*f*) *Water* is an essential item in the diet of most holozoic organisms (there are a small number of desert animals that use metabolic water from respiration) and is necessary for a number of functions. These include the following:

(*i*) Provision of the *continuous phase of living protoplasm* which allows normal activity.

(*ii*) As a *solvent* allowing *transport* of molecules around the body.

(*iii*) A means of *elimination of soluble waste* products.

(*iv*) *Temperature control* in homiothermic animals.

3. The means of collecting food found among holozoic organisms.
Any scheme of classification of such diversity as is to be found in
the ways in which food is collected by animals is bound to be incomplete. One possible method of classification is as follows:

(a) Feeders on large pieces of food:
  (i) *Herbivores*, feeding on plants.
  (ii) *Carnivores*, feeding on animals.
  (iii) *Omnivores*, feeding on both plants and animals.
(b) Feeders on small pieces of food or liquids:
  (i) *Filter feeders* strain off small particles of food from the
environment by bristles, cilia and/or mucus. These include
aquatic organisms as both fresh and sea water contain a large
number of nutritious particles in suspension. Animals such as
whales are also filter feeders although in this case the "krill"
filtered may be crustaceans, each several centimetres long.
  (ii) *Deposit feeders* collect nutritious particles from the substratum, or like earthworms they may ingest large volumes of
the environmental material and digest out organic matter from it.
  (iii) *Fluid feeders* are animals which have some means of
piercing the skin or outer covering of their prey and sucking out
blood or other fluids. Insects, such as aphids, and spiders feed in
this manner.

**4. Examples of the different forms of holozoic feeding from the
major phyla.** *See* Table IX.

TABLE IX.   DIFFERENT FORMS OF HOLOZOIC FEEDING IN THE MAJOR
PHYLA

Phylum or class	Large Pieces of Food			Small Particles or Liquid		
	Herbivore	Carnivore	Omnivore	Filter feeder	Deposit feeder	Liquid feeder
Mammalia	Cow	Cat	Man	Whale		
Aves (birds)	Finch	Hawk	Rook	Duck		
Lower Chordata				Amphioxus	Balano-glossus	
Mollusca	Snail	Squid		Mussel	Teredo	Sea slug
Echino-dermata	Sea urchin	Starfish		Sea lily	Sea cucumber	
Crustacea		Squilla	Crayfish	Barnacle	Shrimp	
Insecta	Butterfly	Mantid	Cockroach	Mosquito larva	Caddis larva	Aphid
Annelida		Rag worm		Fan worm	Lug worm	
Protozoa			Amoeba	Paramecium		

Table IX shows only a small selection of animals but it serves to illustrate the fact that adaptive radiation into all sorts of methods of feeding has taken place independently in many phyla.

NOTE: Following the emphasis given in the new syllabuses to the mammal, the feeding and digestion of these animals will be considered in great detail. As far as other types of holozoic feeding mechanisms are concerned the syllabuses are deliberately open ended. The method followed in this HANDBOOK will be to consider a number of animals representative of the various methods of feeding so that an idea of the sort of adaptations involved may be gained. It should, however, be quite clear to the student that other examples than those selected would be equally suitable. Reference to the questions set on this part of the syllabus indicate the sort of principles that are required to be understood.

FEEDING AND DIGESTION IN MAMMALS

5. Teeth and the mechanical breakdown of food. Mammals are the only class of vertebrates with differentiated teeth. The dentition is closely related to the diet as is the musculature and mechanics of the jaws.

(a) *Herbivores* have incisors which are chisel shaped and adapted for cropping vegetation or, as in the cow, the upper incisors are replaced by a horny pad. There are no canines but instead there is a gap called the *diastema* which allows for more extensive mixing of food in the buccal cavity. The premolars and molars are very similar and have a pattern of complex cusps made of enamel, dentine and cement. These all wear down at different rates and, as the tooth grows continuously, an efficient grinding surface is formed between upper and lower teeth.

A large masseter muscle in the cheek of herbivores is attached to the lower jaw and the loose joint of the latter allows it to move backwards and forwards or side to side.

The dental formula of a sheep is $\dfrac{0033}{3133}$ and for a rabbit $\dfrac{2033}{1023}$.

(b) *Carnivores* have incisors suitable for nibbling meat off bones and grooming. The canines are very large and adapted for the stabbing of prey. Premolars and molars have sharp ridges running along the line of the jaw, and are specialised for tearing meat. The fourth premolar of the top jaw and the first molar of the bottom jaw are specialised for crunching bones and are called the carnassials.

Unlike herbivores the lower jaw of the carnivore is very strong, so too is the force of the bite produced by masseter and posterior temporal muscles. The joint is a tight roller which prevents dislocation of the jaw when chewing tough food.

The dental formula of the dog is $\dfrac{3142}{3143}$.

(c) *Omnivores* have teeth which are relatively unspecialised. A good example is man who has a dental formula $\dfrac{2123}{2123}$. Pigs and bears are also omnivores.

6. The general organisation of the gut.

The alimentary canal consists of a tube with an inner circular and an outer longitudinal set of muscles which provide the peristaltic movement for the passage of the food. Inside the muscle layers are the mucosa and sub-mucosa which enclose the lumen of the gut. The mucosa is glandular and has a very large surface area. Between the mucosa and sub-mucosa is the muscularis mucosa.

The mammalian gut is organised for *extracellular digestion* and is specialised into regions for storage, digestion and assimilation. Besides the secretory cells in its walls the larger glandular masses of the salivary glands, liver and pancreas are also associated with its functioning.

7. Regions of the alimentary canal.

(a) *The buccal cavity.* This is provided with a powerful tongue which, together with the teeth, assist in the mechanical breakdown of food. Salivary glands only produce enzymes in a few mammals such as man but in all cases secrete copious mucus.

(b) *The stomach.* This is divided into a glandular fundic part and a upper cardiac part. There are very large numbers of gastric pits in the former (*see* Fig. 67).

In ruminants a large pouch called the *rumen* is found at the end of the oesophagus. This acts as a temporary store and fermentation chamber for the food which may be passed back into the buccal cavity for further chewing. Within the rumen a large population of bacteria and protozoa exist and these break down

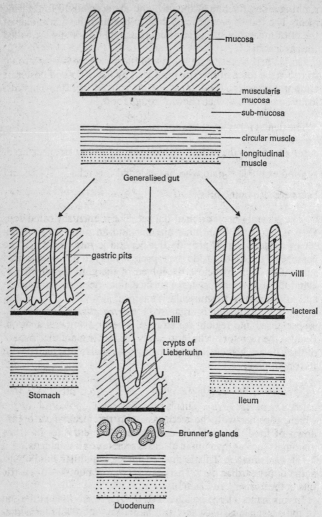

FIG. 67 *A generalised section of the mammalian alimentary canal and the variations found in different regions.*

indigestible substances such as cellulose. Major products of their activity are fatty acids and ammonium ions both of which are assimilated directly. In the liver of the ruminant the ammonia is combined with CO_2 to make urea, and this is later secreted by the salivary glands thus finding its way back to the rumen. Urea is a very good nutrient for the micro-organisms in the rumen, from whose cytoplasm the animal will eventually obtain protein. The protozoa also provide a source of protein, being digested in the stomach of the animal (the abomasum). Between the rumen and the abomasum is the filtering apparatus of the reticulum and omasum.

At the end of the mammalian stomach is the pyloric sphincter.

(c) *The small intestine*. This part is divided into a digestive region, the *duodenum*, and a region of assimilation called the *ileum*. Its surface is greatly increased by the presence of *villi* about one mm in length, and recently electron microphotographs have shown the presence of *microvilli* at the surface of the villa (*see* Fig. 67). It has been calculated that the human intestine has a surface area of 2×10^6 cm^2.

(d) *The large intestine*. At the end of the ileum is the ileocolic valve which helps retain food in the small intestine. The sacculus rotundus also acts as a valve system that can deflect food into the diverticulum of the caecum and appendix.

This latter system is particularly enlarged in herbivores such as rabbits and horses. Like the rumen it also contains a large bacterial and protozoan population which helps in the breakdown of cellulose and synthesis of amino-acids and vitamins. However the gut has no digestive enzymes of its own at this level which is also beyond the main region of assimilation. Thus the caecum–appendix complex is a good deal less effective than the rumen.

The colon, or large intestine, is mainly concerned with the *absorption of water* and storage of faeces.

8. The co-ordination of the digestive system. This is brought about by the activity of both hormones and the autonomic nervous system, reinforced by vagus connections which stimulate the process. A scheme showing the main features of hormone co-ordination is given below in Fig. 68. The autonomic system (*see* XI, 8) has reciprocal innervation to the gut. The parasympathetic connections, stimulated by acetylcholine and mainly carried in the branches of the vagus, tend to increase peristalsis and

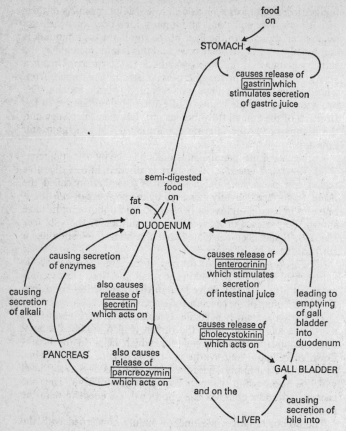

FIG. 68 *The hormonal co-ordination of digestion.*

Some authorities have recently suggested that cholecystokinin and pancreozymin are one and the same hormone, which they term CCK–PZ.

enzyme secretion. The sympathetic connections, which come from segmental ganglia near the spinal cord, and are stimulated by adrenalin, depress digestive activity.

9. The digestive enzymes of mammals. The many enzymes secreted in the different parts of the alimentary canal are best considered in the context of the class of food they hydrolyse. It should be remembered that a part of the specialisation of the gut has been the separation of regions with different pH values associated with the optima for particular enzymes (for details of hydrolytic enzymes *see* IV, 14).

(*a*) *Carbohydrases:*

(*i*) Polysaccharases: salivary amylases and pancreatic amylases hydrolyse starch to maltose.

(*ii*) Glycosidases: duodenal and pancreatic maltase hydrolyse maltose to glucose;

duodenal sucrase (invertase) hydrolyses sucrose to glucose and fructose;

duodenal lactase hydrolyses lactose to glucose and galactose.

(*b*) *Lipases:* pancreatic and duodenal lipase hydrolyse fats to di- and monoglycerides, fatty acids and glycerol.

(*c*) *Proteases* (*peptidases*):

(*i*) Stomach endopeptidase, pepsinogen, is hydrolysed to pepsin. Pepsin hydrolyses petide linkages next to aromatic groups.

(*ii*) Pancreatic chymotrypsin acts in a similar way to pepsin. It is also an endopeptidase.

(*iii*) Pancreatic trypsin, a further endopeptidase, hydrolyses peptide links next to the amino-acids arginine or lysine. It is activated by duodenal enterokinase.

(*iv*) Pancreatic carboxypeptidases hydrolyse peptide links next to —COOH groups at the end of proteins or peptides. This is an exopeptidase enzyme.

(*v*) Intestinal aminopeptidase, an exopeptidase, hydrolyses peptide linkages next to the —NH_2 end of the molecule.

(*vi*) Intestinal dipeptidases hydrolyse dipeptides to amino-acids.

(*d*) *Nucleases:* Nucleases, related nucleotidases and nucleosidases from the intestine and pancreas, hydrolyse nucleic acids into their basic components (*see* IV, 10) which can then be assimilated.

It is thought that the activity of the above hydrolytic enzymes is assisted by the "contact digestion effect" whereby the surface mucus of the gut provides a catalytic surface which facilitates chemical activity.

Some five litres of mucus are produced by the human alimentary canal each day although much of this is reabsorbed in the colon.

10. Assimilation of the products of digestion. Some assimilation of soluble materials takes place in the upper regions of the gut but the bulk of digested substances are assimilated in the villi of the ileum which are provided with a capillary network and the lacteals of the lymphatic system. The special case of fatty-acid uptake from the rumen has already been mentioned.

(a) *Carbohydrates.* These are assimilated as monosaccharide sugars. They are taken up actively against the concentration gradient, probably by a process involving phosphorylation. The surfaces of the villi are rich in phosphorylases. The monosaccharides pass into the capillaries of the hepatic portal system.

(b) *Proteins.* These are assimilated mainly as amino-acids but to a lesser extent as dipeptides. The latter are hydrolysed within the cells of the villi. As with monosaccharides the assimilation is an active process and is blocked by respiratory inhibitors.

(c) *Fats.* The complex of glycerides, fatty acids and glycerol produced as a result of the action of lipase and bile salts on fats forms only one quarter of the amount of fat entering the ileum. This complex acts with unaltered fat to form tiny droplets called *chylomicrons*, some 5·0 nm in diameter. The chylomicrons are taken into the villi pinocytically and there combine together to form larger droplets. By the time these droplets enter the lacteals, or to a lesser extent, the capillaries, they may be up to 1 μm in diameter.

(d) *Vitamins.* These are absorbed directly in the small intestine, although the larger ones are absorbed rather slowly. Fat-soluble vitamins K and D require the presence of bile salts for their assimilation. Vitamin B_{12} is a special case being absorbed from the stomach, provided an "intrinsic factor" secreted by the body is present. Mineral ions are absorbed from the intestine according to the needs of the body. Their uptake is partly under the control of mineralocorticoid hormones. Vitamin D is necessary for the absorption of Ca^{++} and PO_4'''.

11. Fate of the assimilated food substances in the mammalian body. The hepatic portal vein leads to the liver and monosaccharide sugars and amino-acids are transported by this organ. The hepatic portal vein sub-divides to supply the great network of sinusoids in the liver lobules. Monosaccharides are taken up by the cells of the sinusoids and condensed to glucogen which is stored. The hormone insulin encourages this process while adrenalin brings about the release of sugar from the liver.

Amino-acids are also taken up by the cells of the liver sinusoids. Those that are required by the body are returned to the circulation and others, or any excess, are deaminated with the removal of ammonia. The resulting keto-acids can be converted to glycogen via the Krebs cycle, or respired. The ammonia can be used to form other types of amino-acids or else it may combine with CO_2 to give urea which is excreted. The combination is performed via the ornithine cycle (*see* IX, **18**).

Fats pass via the lymphatic system to the cisterna chyli and thence to the thoracic ducts at the base of the jugular veins. Some will be stored in the body and others taken up by the liver where they can be respired (*see* VIII, **7**). Fats are also used in the synthesis of essential parts of the cell.

SELECTED EXAMPLES OF INVERTEBRATE FEEDING AND DIGESTION

12. Feeding and digestion of cockroaches. Cockroaches are omnivorous and feed on solids, i.e. any scraps that they are able to find. They start with food that is in large pieces and it has to be broken down to particles small enough to swallow.

(*a*) *The mouthparts.* These are derived from segmental appendages which have been modified in the evolution of a specialised head. The palps are sensory and allow sampling of the food which they push against the grinding edges of the mandibles and maxillae. The labium acts as a lower lip (*see* Fig. 69). During chewing saliva is poured over the food. This contains amylase which commences the digestion of starch.

(*b*) *Digestion in the crop.* The semi-liquid food mass is passed into the large crop which is lined with chitin. Into the crop digestive enzymes of all classes are passed backwards from the secretory cells of the mesenteric caeca. Digestion proceeds in the crop until the food is reduced to soluble form or very small particles. This activity is assisted by the teeth of the gizzard which also act as a filter.

(*c*) *Assimilation in the midgut.* The digested food enters the midgut and mesenteric caeca where it is assimilated by the same cells that produce the digestive enzymes. After assimilation, the food is transported across the haemocoel by the blood and either enters the cells of the body or is stored in the fat of the body.

(*d*) *The hind gut and rectum.* These take up water from the faeces which are excreted in solid form.

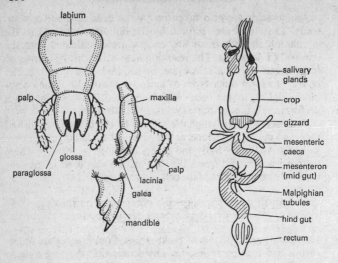

FIG. 69 *The mouthparts and alimentary canal of the cockroach.*

NOTE: Insects do not use mucus to line their digestive tracts but instead secrete a *peritrophic membrane* of which the main function seems to be to protect the gut walls from abrasion by hard food.

13. The capture and digestions of food by Hydra. This animal would also be classed as a feeder on solid foods and, in fact, is able to deal with prey almost the same size as itself.

(*a*) *Capture of food.* The extended hydra has long tentacles which provide an extensive net for the capture of the small crustaceans on which the animal feeds. A peptide substance called *glutathione* is released from the prey and this triggers off a series of feeding responses in hydra. One of the actions of this response is the gaping of the mouth and increased activity of the tentacles. The feeding reflex takes thirty seconds to begin and lasts for approximately half an hour. An internal control system takes many hours to build up before the mechanism can operate again.

On the tentacles are batteries of nematocysts. When the sensitive cnidocils of these are touched by the prey, muscular and osmotic changes lead to the discharge of a fine thread, of which there are three kinds. The end of this thread may be sticky (glutinant) or it may wrap itself around the prey (volvont), or it may contain a toxin (penetrant) which will paralyse the prey. Nematocysts are independent effectors but their threshold for firing depends on the nutritional state of the whole animal.

(b) *Passage of food to the mouth.* According to the amount of stimulation received a varying number of muscular epithelial cells are brought into action. Small prey may be taken by a single tentacle whereas larger organisms will require co-operation of all the tentacles for their capture. The prey, paralysed and attached to the tentacles, is brought into the mouth.

(c) *Digestion in the enteron.* Glandular cells of the endothelium pour out digestive enzymes which bring about partial breakdown of the food. Hydra tends to localise digestive activity in one part of the enteron at a time by contracting horizontally and forming a "pouch". The products of this extracellular digestion are then wafted around in the enteron by cilia and soluble substances are taken up by digestive cells. These latter also have amoeboid processes and are able to take up quite large pieces of food and complete their digestion intracellularly. Waste products of digestion are returned from these cells to the enteron and are passed out through the mouth during one of the periodic emptyings of the enteron.

It should be noted that this mixture of extra- and intracellular digestion is primitive.

Besides feeding as a carnivore, hydra receive a good deal of nutrients from the symbiotic zoochlorellae of the endoderm. Experiments on normal and "white" (containing no algae) hydra show that the powers of the latter to withstand starvation are very much reduced.

14. Feeding and digestion in earthworms. These animals feed on small particles of nutritious matter in the soil. As their food has to be mechanically broken up and yet is largely in particle form earthworms come between the two main classes of holozoic feeders. The process of feeding is as follows:

(a) Soil and leaves are taken in by the mouth and enter the *buccal cavity*.

(b) The food then moves to the *pharynx* where a protease secreted from salivary glands starts the process of digestion. Calcium carbonate is added to the food by the calcareous glands. The food is stored in the crop.

(c) The *gizzard* is muscular and grinds the food and more protease is added from glands. The food in a semi-liquid state enters the intestine.

(d) *The intestine* runs from segment 19 to the end of the worm. Its surface is much increased by the typhlosole and from cells in its wall lipase, amylase, protease, chitinase and cellulase are secreted. This is a very broad complex of enzymes and the worm is unusual in its possession of cellulase. The enzymes are very suitable to hydrolyse the rather indigestible substances found in soil.

It should be noted that the feeding activity of earthworms does much to increase the fertility of soil by aerating it and by turning it over it accelerates the speed at which organic matter can be converted to humus.

15. The feeding mechanism of Sabella.

This animal is a tube worm and a polychaete. It is found with related species in huge colonies on the sea bed feeding on the tiny particles that sink down from the surface. Sabella is a good example of a particulate feeder that uses ciliary currents to move its food to its mouth.

The worm has a circle of long tentacles which are covered with ciliated pinnae. The beating of the cilia draw currents of water through the tentacles which filter out any food particles in the water. Further ciliary action moves the particles to the groove that runs along the main arm of the tentacle and once in this groove they are transported towards the centrally-placed mouth.

A complex sorting apparatus at the base of the tentacles sorts the particles into various sizes. The large ones tend to be rejected, the medium ones become incorporated into the tube of the animal and are cemented there by secretions, while the smallest particles are ingested.

16. Liquid-feeding mechanism of aphids.

Strictly according to the classification that has been adopted, aphids should be considered as parasites. It will, however, be appropriate to discuss them as examples of animals that are specially adapted for feeding on liquids as their mouthparts indicate the sort of modifications required.

The labium contains the other mouthparts and is jointed so that it can act as a guide and support to the piercing tubular mouthparts as they are pushed into the plant. The two mandibles are chisel shaped and together with the fused maxillae that they enclose assist piercing through the tissues of the epidermis and cortex. There are two tubes in the maxillae, the anterior being for the passage of saliva which, as with many other liquid feeders, contains an enzymic anticoagulant, while the posterior tube lead into a muscular pharynx which provides the suction for the uptake of sugars from the phloem. Excised aphid heads have been used to sample phloem juices. Aphids have to take up a great deal of liquid from the phloem to obtain enough protein and the excess sugars are excreted via the tubercules at the end of the abdomen.

17. Filter feeding in Paramecium.

The ciliate *Paramecium* takes up bacteria and other living and non-living matter from the fresh water in which it lives. The feeding process is as below:

(*a*) The body of Paramecium is covered with cilia and these are particularly large along the oral groove. When the animal is stationary or slowly moving the beating of these cilia creates a vortex current and particles of food are wafted into the cytopharynx. Rows of cilia in the cytopharynx cause these particles to move to the oesophageal sac where a food vacuole forms. It is possible for the animal to distinguish and reject unsuitable material.

(*b*) The food vacuole follows a complicated course around the body of the animal. At first it decreases markedly in size and at the same time becomes very acidic, the *p*H falling near 1. Any living organisms within the vacuole are killed during the acidic phase. The vacuole then increases in size and its contents become alkaline. Enzymes of the trypsin type together with carbohydrases and lipases are secreted into the vacuole and the digestion of the food commences.

(*c*) Digestible material is hydrolysed and absorbed, proteins probably in the form of peptides, and an indigestible residue remains. This residue is lost from the anal pore of which the position is constant.

NOTE: A similar cycle of events takes place in the food vacuoles of *Amoeba*. During the assimilation phase the wall of the vacuole develops many filamentous protrusions and the semi-digested food is taken up by pinocytosis. The vacuole is digested in less than one hour. There is a regular twenty-hour period of feeding and growth at 23°C which is followed by longitudinal fission.

THE NUTRITION OF SAPROPHYTES

18. General features of saprophytes related to their mode of nutrition. Saprophytes, as we have seen, are organisms that cannot synthesise their nutritional requirements wholly from inorganic sources and which take in dissolved substances all over their body. The majority of saprophytes particularly fungi are associated with the decay or breakdown of plant or animal material and in the form of bacteria they are found in great quantities in soil and on the sea bed. Most saprophytes have a number of features in common related to their means of nutrition. They are as follows:

(*a*) The "host" of the saprophyte is likely to be randomly distributed in the environment in the form of a dead plant or animal body or excreta, etc. For this reason the saprophyte will tend to have the following features:

(*i*) *An effective means of dispersal* to colonise new material. This might take the form of spores with a fruiting body, the sporangium, growing up out of the host. Such fruiting bodies are clearly seen in *Mucor* and the mushroom.

(*ii*) *A very large number of sexually or asexually-produced offspring* to make up for the inevitable loss involved in colonising new material.

(*iii*) *Resistant spores or zygospores* which will allow the organism to survive unfavourable periods.

(*b*) By its nature the material or host on which the saprophyte feeds is likely to have a very temporary existence. Related to this the saprophyte will tend to show certain characteristics:

(*i*) Very *rapid growth* to colonise the material quickly before it becomes desiccated or eaten by other organisms. The taking in of food all over the body surface naturally increases the speed of growth.

(*ii*) An *enzyme complex* suitable to the material on which the saprophyte feeds. Both bacteria and fungi produce a very wide range of hydrolytic enzymes and some can break down substrates such as lignin, cellulose and chitin.

(*iii*) The ability to respire *anaerobically* as the supply of oxygen rapidly becomes depleted in a semi-liquid food mass. Yeast is able to do this as are many other saprophytic fungi such as Mucor. The same is true of many soil bacteria and there are a few species of bacteria that can only survive in the absence of oxygen.

There are a great many saprophytes and a very limited supply of nutrient material so that competition is very intensive. Associated with this some saprophytic organisms secrete substances which are toxic to others. The best known cases of this are among soil fungi such as *Penicillium*, and the purified extracts from these fungi have been of great importance in combating bacterial infections in man and some of his domestic animals.

19. Yeast as an example of saprophytic nutrition. There are many species of *Saccharomyces* or yeasts and a few of these are parasitic and cause diseases in man and other animals. The majority of yeasts live as spores in the soil and are blown on to ripening fleshy fruits with the dust, or else contaminate the fruit if it drops. Yeasts cannot attack healthy fruits but easily invade bruised and dead tissues.

(*a*) *Uptake of nutrients.* The yeast cells secrete carbohydrases on to the food material and conversion of disaccharides to monosaccharides occurs. Autolysis will also take place in the dead cells of the fruit making insoluble carbohydrates available. Mineral ions are taken up selectively by the yeast cells which rapidly increase in size. Under favourable conditions budding may occur several times an hour and large populations are built up. Stored carbohydrate is found in the cells in the form of glycogen.

(*b*) *Respiration.* The metabolism of yeast functions most efficiently in aerobic conditions but part of its adaptation to its saprophytic mode of life is its ability to respire anaerobically and to tolerate concentrations of ethanol. The pathway of anaerobic respiration is as follows:

glucose $C_6H_{12}O_6 \rightarrow 2CH_3CO.COOH$ (pyruvic acid) $+$ energy

decarboxylase enzyme

$$2CH_3CHO \text{ (acetaldehyde)} + 2CO_2$$

$+2H.$

$$2CH_3CH_2OH \text{ (ethanol)}$$

Some strains of yeast are able to survive in concentrations of over 5 per cent ethanol and the vats in breweries are so designed that a period of rapid growth under aerobic conditions is followed by fermentation and alcohol production under anaerobic conditions.

(*c*) *Sexual reproduction.* Yeast is a member of the Ascomycetes group and periodically forms ascospores. These may fuse in pairs to produce a diploid cell. Under unfavourable conditions a resistant spore is formed around either haploid or diploid phases of the fungus.

(*d*) *Success of yeasts.* Yeasts are very widespread and successful saprophytes and this is due to a number of factors:

(*i*) The food substrates they utilise are very common.

(*ii*) They can survive in the soil in spore form for long periods.

(*iii*) They rapidly colonise suitable food and are able to survive in anaerobic conditions and thus compete successfully with other micro-organisms.

THE FEEDING MECHANISM OF PARASITES

20. Definitions. A *parasite* is an organism which feeds on or in another organism, termed the host, and causes it harm. The latter may vary from causing the death of the host, as with the blight on the potato plant, to causing very minor irritations, as with fleas on a dog, but the idea of harm is implicit in a definition of parasitism. Parasitism should be distinguished from other relationships between organisms, such as the following:

(*a*) *Symbiosis,* where two organisms live together in a state of mutual benefit; e.g. the green algae of hydra or the algal and fungal associations of the lichens or the microflora of the ruminants' gut.

(*b*) *Commensalism,* where one organism derives food or shelter from another without causing it actual harm, e.g. the polychaete worm that lives in the shells of hermit crabs or the skuas which dive at other sea birds and cause them to drop fishes which the skua then takes.

21. Examples of parasitic organisms. An animal which feeds on the outside of another is termed an *ectoparasite* while one that lives inside the tissues of its host is termed an *endoparasite*. These terms are not applied to plant parasites.

Both animal and plant parasites may use *vector* organisms such as mosquitoes or flies or aphids to transmit their infective stages from one host to another.

Despite the very large number of parasitic species, and there

are clearly more parasites than free-living organisms, the parasites are mainly drawn from certain phyla or plant groups. These,
together with examples of specific parasites are as below:

(a) *Animal parasites:*
(i) *Protozoa:*
Flagellates, e.g. *Trypanosoma* (sleeping sickness).
Rhizopods, e.g. *Entamoeba* (dysentry).
Sporozoans, all parasites, e.g. *Plasmodium* (malaria).
(ii) *Platyhelminthes:*
Trematodes are all either ecto- or endoparasitic flukes, e.g.
Fasciola (sheep liver fluke).
Cestodes are endoparasites, the tapeworms such as *Taenia*.
(iii) *Nematodes:*
Roundworms parasitise both plants, e.g. *Heterodera*, the eel
worm, and animals, e.g. *Ascaris, Filaria* (elephantiasis), *Necator*
(hookworms).
(iv) *Annelids:*
Hirudinae—the leeches—are mostly ectoparasitic blood-suckers.
(v) *Arthropods:*
Insects—lice, fleas, mosquitoes and aphids, live on the outside of
their hosts and there are very many endoparasites such as the
larvae of *Ichneumon* and *Tabanid* flies which kill other insects.
Arachnids—mites and ticks, are ectoparasites especially of birds
and mammals.
(b) *Plant parasites:*
(i) *Bacteria:* There are very many species of bacteria which
may parasitise animals. Many human diseases such as typhoid
and tuberculosis are due to such organisms.
(ii) *Fungi:* The blights, smuts and rusts, damping off and
powdery mildews are all widespread plant parasites. These are
also some fungi which parasitise animals such as *Tinea* on man
and *Empusa* on flies. The aquatic fungus *Saprolegnia* is a parasite
of fishes.
(c) *Viruses:* Viruses are all parasitic. Many animal diseases,
such as poliomyelitis, foot-and-mouth and foul brood which
attacks bees, are caused by viruses. There are also numerous
plant diseases such as scabs and mosaics due to virus infections.

22. Problems of parasitism.

Some of the points raised in **13** in
connection with the problems of saprophytes are equally relevant
here. As in the case of saprophytes the food, in this case the hosts,
of the parasites is randomly distributed within the environment.

In order to find and invade its particular host (and of course parasites are very much more specific about their hosts than saprophytes because of the difficulties of invading living tissues) the parasites tend to have certain adaptations:

(a) *Large numbers and wide dispersal of offspring* are necessary. In many cases vector organisms are involved so the life cycles of parasites may be very complex.

(b) *Resistant stages* that can withstand unfavourable conditions are needed. The cysts of certain parasitic protozoa are some of the most resistant organisms known.

(c) *A means of invasion* of the host is necessary and thereafter a means of *resisting destruction* by the antibody and enzyme systems of the host.

(d) On the whole, though by no means invariably, *a stage of equilibrium* develops in what is called a "well adapted" parasite. These may weaken or damage their hosts but they do not kill them (as with trypanosomes in wild game). Recent parasitic mutants or newly-introduced parasites may cause almost one hundred per cent deaths of the hosts, e.g. the introduction in the eighteenth century of measles to the South Sea Islands or the introduction of myxomatosis in the 1950s to European rabbits. In this latter case the same build-up of resistance took place that had occurred in the Australian rabbits. A huge initial kill of over 99 per cent was followed by increasing resistance to the disease and the present situation is that of an uneasy equilibrium which sometimes swings towards the rabbits survival and sometimes against it. Clearly it is not to the advantage of a parasite to eliminate its host species.

23. Are parasites degenerate organisms?

Because they live in or on the bodies of other organisms many parasites tend to lose those organs that we associate with free life. In the case of animal parasites these will be largely the locomotion, co-ordination or trophic organs, while higher-plant parasites like dodder lose chlorophyll, leaves and roots. From this point of view the parasite is certainly degenerate.

(a) Conversely, the parasite may have other highly-developed features. The parasite may show extreme adaptations of its reproductive system in the production of vast numbers of offspring (*see* Table X).

(b) There may also be *complex feeding mechanisms* as for

TABLE X.   NUMBER OF EGGS PRODUCED BY SELECTED PARASITES

Species	Number of eggs produced
Diphyllbothrium (fish tapeworm of man)	1,000,000 per day
Taenia solium (pork tapeworm of man)	800,000,000 total egg production
Ascaris (round worm of pig and man)	200,000 eggs per day
Hookworm species	35,000 eggs per day

example the haustoria of many phycomycetes or the piercing and sucking mouthparts of mosquitoes and aphids.

(c) Gut and ectoparasites such as fleas often have *elaborate means of attachment* to the surface of their hosts and the whole body may be flattened in such a way as to prevent easy dislodgement.

In general it is best to consider that the majority of plant and animal parasites are highly-adapted and specialised organisms and not degenerate versions of their free-living relations.

24. Examples of life histories of parasites and the means of overcoming the parasite.

(a) *Phytophthora infestans* (the potato blight). The life cycle of the parasite is shown in Fig. 70.

(i) *Damage caused to host:*

(1) Rapid destruction of the leaves and tissues of the plant above ground partly due to the blight but also to the invasion of other micro-organisms.

(2) Destruction of tubers in the soil and in storage clamps.

(3) Even with modern methods the total loss to farmers is still calculated in millions of pounds per year.

(ii) *Means of controlling the parasite.* This is based largely on the life history and consists of the following measures:

(1) Spraying the crop with fungicides such as the Bordeaux mixture, especially after wet weather.

(2) Broadcasting to farmers the likely times of blight.

(3) Destruction of blighted tubers by burning.

(4) Storage of tubers in well-ventilated clamps.

(5) Rotation of potatoes with other crops not subject to blight.

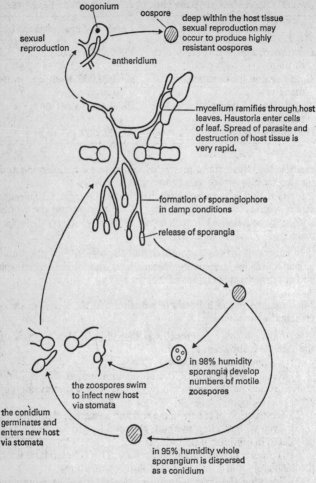

oogonium

oospore

deep within the host tissue sexual reproduction may occur to produce highly resistant oospores

sexual reproduction

antheridium

mycelium ramifies through host leaves. Haustoria enter cells of leaf. Spread of parasite and destruction of host tissue is very rapid.

formation of sporangiophore in damp conditions

release of sporangia

in 98% humidity sporangia develop numbers of motile zoospores

the zoospores swim to infect new host via stomata

the conidium germinates and enters new host via stomata

in 95% humidity whole sporangium is dispersed as a conidium

FIG. 70   *The life cycle of the potato blight,* Phytophthora infestans.

    (6) Treatment of tubers by fungicides before planting.

    (7) Breeding of resistant strains of potatoes.

    (*b*) *The sheep liver fluke.* The life cycle of the parasite is shown in Fig. 71.

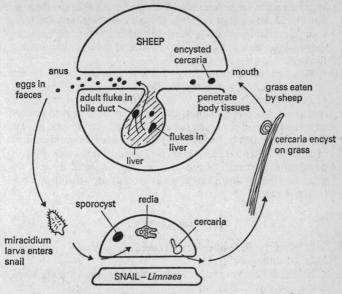

FIG. 71   *The life cycle of* Fasciola hepatica, *the sheep liver fluke.*

(*i*) *Damage caused to the host.* Death of the sheep is caused by general destruction of the liver, particularly by a heavy infestation.

(*ii*) *Control of the parasite.* This involves the following measures:

(1) Destruction of the snail by the introduction of geese and ducks to low-lying pastures.

(2) Draining swampy pastures.

(3) Treatment of fluke infections in the sheep.

(4) Addition of lime to the soil which prevents the hatching of the fluke eggs.

25. Parasites as agents of natural selection. As already mentioned parasite and host tend to evolve an uneasy equilibrium between them and recent work shows that healthy vigorous organisms may produce specific chemicals inimicable to their parasites. An example is phaseolin made by some bean plant species.

Very young and old organisms are particularly susceptible to attack by parasites, as are individuals rendered unhealthy through nutritional deficiencies or other causes. For these, and for individuals that are genetically more susceptible to a given parasite, they may act as a powerful form of natural selection eliminating the weak.

Most species of living things are kept in check partly by the deaths caused by their parasites, directly or indirectly. Since the work of Pasteur and many others, from the mid-nineteenth century to the present times, mankind has been able to understand and control the many parasites that plague him and in past times did so much to limit his population. The successful control of these parasites has had a great deal to do with the great human population explosion of the past century. It seems there is no good controlling death rate if we cannot, at the same time, control birth rate.

PROGRESS TEST 7

1. What is the difference between an essential and non-essential amino-acid? **(2)**

2. Which different methods of feeding are found among holozoic organisms? **(3)**

3. Give the dental formula of the dog. **(5)**

4. What are the parts and functions of the rumen? **(7)**

5. Name four classes of hydrolytic enzymes found in the human gut. **(9)**

6. Which hormones are released from the duodenal wall? **(8)**

7. How are fats assimilated? **(10)**

8. How does the cockroach break up its food? **(12)**

9. What part does glutathione play in the feeding responses of *Hydra*? **(13)**

10. Which regions of the alimentary canal are found in earthworms? **(14)**

11. Why do filter feeders tend to sort out particles by size? **(15)**

12. What sequence of processes takes place in the food vacuoles of *Paramecium*? **(17)**

13. Which problems of survival confront saprophytes? **(18)**

14. How does one define a parasite? **(20)**

15. In what respects might parasites be considered degenerate organisms? **(23)**

16. How does knowledge of the life history of a named parasite lead to effective control measures? **(24)**

EXAMINATION QUESTIONS

The figures in **bold** type indicate the marks allocated to each question or part question.

1. (*a*) In a short account of a **named** internal parasite (e.g. a platyhelminth or nematode) of a vertebrate animal, describe (*i*) its method of nutrition, and (*ii*) any adaptations to parasitism. **(9)**

(*b*) Many parasites spend some periods of their life cycles living independently of their hosts. Discuss briefly the advantages and disadvantages to the parasite of having an independent stage in its life cycle. **(9)**

(Total **18***) Cambridge,* 1977

(*a*) State the term used to describe the mode of nutrition of each of the following organisms.

(*i*) Man.

(*ii*) Nitrifying bacteria.

(*iii*) A green plant.

(*b*) (*i*) Into which category would the nutrition of saprophytes be placed?

(*ii*) Explain your choice.

(Total **6***) part question, Associated Examining Board,* 1977

3. Saprophytic and parasitic fungi have a number of features in common.

(*a*) State *one* feature of their life cycles which they have in common.

(*b*) State *one* feature related to their physiology which they have in common.

(*c*) Relate *each* feature chosen to the biological problems of fungal parasites and saprophytes.

(Total **8***) Associated Examining Board,* 1977

4. Describe what you would expect to happen when a suspension of proteins, fats and starch is hydrolysed by activated pancreatic enzymes in a length of Visking (cellulose) tubing suspended in water in a beaker. **(9)**

To what extent does this experiment represent the structures and functions involved in the processes of digestion and absorption in a mammal? **(9)**

(Total **18***) Cambridge,* 1977

5. In August 1976 the death was reported of a man "who would not eat meat, milk, butter, cheese, eggs or fish but subsisted only on honey, fruits, salads and wheat germ. The coroner brought in a verdict of accidental death from malnutrition due to self-neglect." (*Daily Telegraph* report.)

How would you have explained to the coroner the most likely reasons for the man's death?

(Total 20) Associated Examining Board, 1977

6. Describe the main ways in which mammalian herbivores are adapted for the ingestion and digestion of plant material (answers may deal with one or more species). **(20)**

Oxford and Cambridge, 1977

7. Amplify the following: "The contact of chyme from the stomach with the cells lining the duodenum sets in motion a series of important events".

after Oxford and Cambridge, 1974

8. Discuss the ways in which parasites cause biological damage and economic loss.

Oxford and Cambridge, 1975

9. Using examples to illustrate your answers show how knowledge of the history of parasites may be used in the control of them. **(20)**

Oxford and Cambridge, 1977

10. Relate the feeding methods of (*a*) sedentary and (*b*) free moving animals to their mode of life.

London, 1975

11. What is a parasite? Describe the life cycle of a *named* parasite. Discuss the host-parasite relationships in the example you have selected. **(4, 9, 7)**

London, 1976

12. Describe how a *named* mammal detects, ingests and masticates its food. Explain why thorough mastication is of more importance to a herbivore than to a carnivore. **(20, 5)**

London, 1976

CHAPTER VIII

Respiration

TISSUE RESPIRATION

1. Definition of respiration. Respiration is an essential activity of all living things by which they obtain energy for all other metabolic processes.

NOTE: Although respiration is often confused with breathing, or gaseous exchanges, it is important to realise that while uptake, transport and utilisation of oxygen are often associated with respiration they need not be involved. Many organisms respire anaerobically, that is without use of oxygen, either for limited periods or for their whole lifetime, e.g. botulin bacteria.

In its true physiological sense respiration is a chemical activity taking place within the protoplasm of the cell. It consists of the downgrading of the respiratory substrate, often glucose, into smaller molecules which are released as waste. During this downgrading process the chemical energy contained in the original substrate is partly transferred to energy-storing compounds in the cell and partly lost as heat.

2. Energy carriers and stores in the cell. One of the most important of these energy carriers and stores is adenosine triphosphate (ATP). The energy is carried in the third phosphate bond which on hydrolysis yields adenosine diphosphate (ADP), phosphate and 33 kilojoules.

$$ATP \rightleftharpoons ADP + PO_4 + c.\ 33 \text{ kilojoules per molecule}$$

The high-energy phosphate bonds may also be transferred to and stored in other substances; thus in vertebrate muscle creatine is converted to creatine phosphate and in some invertebrates arginine is converted to arginine phosphate. This releases the ADP to collect further energy-rich phosphates.

High-energy substances such as ATP are involved in the synthesis of complex molecules by the cell. They are important in the provision of energy for active intake and secretion across membranes against diffusion gradients. Muscle contraction and nerve conduction depend on ATP and in all the reactions in which it takes part heat is liberated. This heat usually represents an *unavoidable loss of energy* to the living system but in some homiothermic animals it is conserved.

THE CHEMICAL PATHWAYS OF RESPIRATION

3. Respiration of carbohydrates. This takes place at first anaerobically and then, when oxygen is present, aerobically. If no oxygen is present, complete anaerobic breakdown may occur. The parts are described in **4, 5, 6.**

4. Anaerobic glycolysis. For most organisms *carbohydrates* are the main respiratory substrates. Either as poly-, oligo- or monosaccharides (*see* IV, 3), they may enter downgrading pathways. It is simplest to think of them starting as glucose but larger units may be phosphorylated directly and do not need to be turned to actual glucose molecules before being respired.

All organisms have a common chemical pathway which makes up the first part of the respiratory sequence of reactions. Since this pathway does not use up oxygen and often involves sugars it is termed *anaerobic glycolysis* (*see* Fig. 72). At the end of anaerobic glycolysis there has been a net energy gain to the cell of 2ATP and the hexose sugar or other carbohydrate involved is now in the form of pyruvic acid.

It is characteristic of the various enzymes involved in the process that they are contained in the cell matrix and not in the mitochondria.

In plants the pyruvic acid passes directly into the Krebs cycle if oxygen is present above a critical concentration (*see* **5**). If not it may be converted to alcohol, etc. (*see* **6**). In animals such as ourselves, excess pyruvic acid is reduced to lactic acid, $CH_3CH(OH).COOH$, which is returned to the liver for the resynthesis of carbohydrate.

5. Aerobic processes leading to the main release of energy. The pyruvic acid from anaerobic glycolysis reacts with *coenzyme A* so that the acetyl group $CH_3CO—$ becomes attached giving *acetyl-co A* and carbon dioxide is liberated. Acetyl-co A enters the Krebs cycle (also called the *citric acid* or *tricarboxylic acid cycle*) and undergoes a series of changes which may be shown as in Fig. 73.

It will be seen that at various points in the cycle carbon dioxide

CHEMICAL PATHWAY   COMMENT

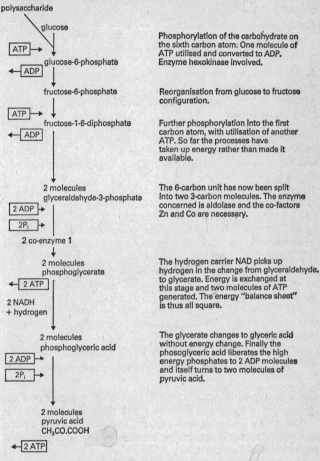

polysaccharide

glucose

ATP →

glucose-6-phosphate

← ADP

Phosphorylation of the carbohydrate on the sixth carbon atom: One molecule of ATP utilised and converted to ADP. Enzyme hexokinase involved.

fructose-6-phosphate

Reorganisation from glucose to fructose configuration.

ATP →

fructose-1-6-diphosphate

← ADP

Further phosphorylation into the first carbon atom, with utilisation of another ATP. So far the processes have taken up energy rather than made it available.

2 molecules
glyceraldehyde-3-phosphate

2 ADP →

2 Pᵢ

The 6-carbon unit has now been split into two 3-carbon molecules. The enzyme concerned is aldolase and the co-factors Zn and Co are necessary.

2 co-enzyme 1

2 molecules
phosphoglycerate

← 2 ATP

2 NADH
+ hydrogen

The hydrogen carrier NAD picks up hydrogen in the change from glyceraldehyde. to glycerate. Energy is exchanged at this stage and two molecules of ATP generated. The energy "balance sheet" is thus all square.

2 molecules
phosphoglyceric acid

2 ADP →

2 Pᵢ

The glycerate changes to glyceric acid without energy change. Finally the phosoglyceric acid liberates the high energy phosphates to 2 ADP molecules and itself turns to two molecules of pyruvic acid.

2 molecules
pyruvic acid
$CH_3CO.COOH$

← 2 ATP

FIG. 72 *The stages of anaerobic glycolysis.*

and hydrogen are released. While the carbon dioxide is released the hydrogen passes through a number of further stages until it eventually combines with oxygen to form water. This vital part of chemical respiration is done by means of *hydrogen acceptors,*

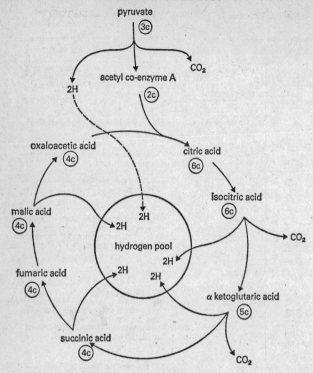

FIG. 73   *The Krebs or TCA cycle.*

The number of carbon atoms in each substrate are ringed.

which receive hydrogen from a substrate in the presence of a dehydrogenase enzyme and oxidases which bring about the combination of the hydrogen and oxygen. The nomenclature of the hydrogen acceptors is somewhat confusing as various synonyms have been used (*see* Table XI).

The path of the hydrogen from the Krebs cycle is shown in Fig. 74. The terminal oxidase is usually cytochrome, an iron-containing enzyme. This enzyme is poisoned by cyanide, which accounts for the toxic effect of the chemical on aerobic respiration.

During the stages of the Krebs cycle and subsequent formation

TABLE XI. NOMENCLATURE OF HYDROGEN ACCEPTORS IN
RESPIRATION

General name	Previous name	Present name
Co-enzyme 1	Diphosphopyridine nucleotide (DPN)	Nicotinamide adenine dinucleotide (NAD)
Co-enzyme 11	Triphosphopyridine nucleotide (TPN)	Nicotinamide adenine dinucleotide phosphate (NADP)
—	flavo adenine dinucleotide (FADN)	—

of water a number of ATP molecules are generated from ADP
and phosphate. These form at those points in the cycle where
hydrogen is produced and, for each molecule of the original
pyruvic acid, yield 15ATP. A hexose sugar gives therefore
2×15ATP (two molecules of pyruvic acid go through the cycle),
2ATP from anaerobic glycolysis and a further 6ATP from the
reduced NAD produced in the change from glycerate to glyceric
acid (see Fig 72).

A total yield of energy from a molecule of hexose is thus
38ATP, which is some sixty per cent of the theoretical energy
content. Energy exchanges in living organisms compare very
favourably with those of machines such as steam engines where
only about five per cent of the energy of the fuel is converted to
work.

6. Relation of enzymes to mitochondrial structure. It now seems
clear that the enzymes involved in anaerobic glycolysis lie outside the mitochondria. Yeast, for example, is able to carry out
anaerobic ATP formation without mitochondria present.

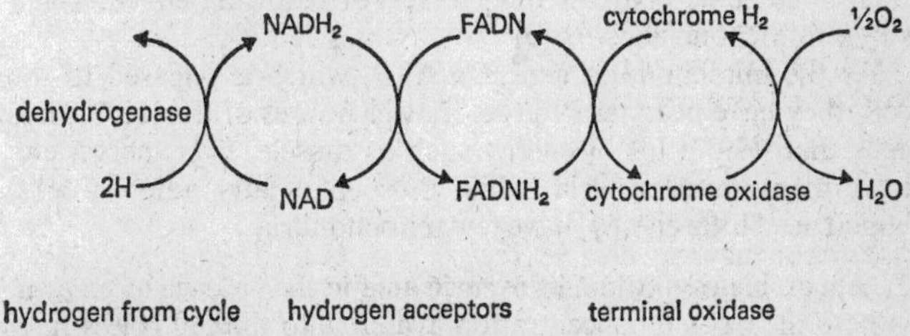

FIG. 74 *The path of hydrogen after the Krebs cycle.*

Normally, therefore, acetyl co-A will enter the mitochondria from the surrounding cytoplasm and pass into the lumen. Here the various enzymes of the tri-carboxylic cycle are either in solution in the matrix or else possibly loosely bound at the inner membranes of the cristae. We do know for certain that the flavoproteins and the cytochromes are situated on the inner mitochondrial membranes and that the final generation of ATP takes place by the enzyme activity of the ATP synthetase molecules on the fundamental particles. Once ATP has been generated it can readily diffuse out into the cell cytoplasm where it is required. A summary scheme could be represented as in Fig. 75.

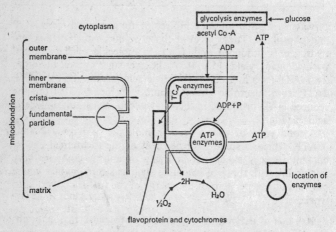

FIG. 75 *Location of enzymes in the mitochondrion.*

Most cells will contain many thousands of mitochondria and there may be as many as 100,000 sets of oxidising enzymes in a single mitochondrion.

As the mitochondria generate ATP, which is released to the cell, they have been termed the "power houses of the cell". Thus cells that need a lot of energy such as muscle, liver, nerve, etc. have many mitochondria, while those like fatty adipose cells, which need little energy, have few mitochondria.

7. Anaerobic respiration of pyruvic acid in the absence of oxygen.
Where the oxygen concentration available to the cell is less than two per cent the pyruvic acid formed at the end of anaerobic

glycolysis may proceed to acetaldehyde by the loss of carbon dioxide. The reduced NAD formed in glycolysis can then transfer its hydrogen to the acetaldehyde forming ethyl alcohol, i.e.

$$CH_3CO.COOH \rightarrow CH_3CHO + CO_2 \rightarrow C_2H_5OH + 2H$$

No further energy is released by these reactions so that the net gain to the cell is only the 2ATP from the anaerobic glycolysis pathway.

NOTE: This type of respiration, also called fermentation, is very inefficient and is found in few organisms. It is found in gut parasites such as tapeworms which live in an environment of low-oxygen content. On the whole plants, having much lower energy demands than animals, can withstand periods of anaerobic respiration much more readily than animals. Certain fungi, e.g. yeast, may be very tolerant of anaerobic conditions and can maintain life in up to concentrations of twelve per cent alcohol. This gives them the ability to compete successfully for food in the environments in which they live.

8. Other energy paths. Besides the well-known pathway of glucose breakdown described, other common but generally less well known paths of energy exchange occur. One of these is the pentose path involving five carbon sugars. In all these paths ATP is by no means the only high energy product; guanine triphosphate and uracil triphosphate may also be formed.

9. Respiration of fats. Fats and related compounds that are stored in the cell are hydrolysed into glycerol and fatty acids before respiration. The fatty acids react in such a way with co-enzyme A that an acetyl group is removed and a smaller fatty acid left. This is a process of oxidation. The acetylco-A so produced is passed into the Krebs cycle and undergoes the changes described in **5**. Consecutive acetyl groups are removed until the fatty acid has been completely oxidised.

The enzymes for the oxidation of fats are in the mitochondria, and it should be noted that a 6-carbon fatty acid yields 44ATP on oxidation. Fats are a richer source of energy than carbohydrates but the continued "rotation" of the Krebs cycle demands that a supply of carbohydrates be passing through the respiration path. This has been expressed in the notion that "fats burn in the flame of carbohydrates".

Under anaerobic conditions the partial respiration of fats leads to the accumulation of toxic products such as acetone.

10. Respiration of proteins. Proteins are first broken down to amino-acids and these are subsequently deminated leaving the CHO as a keto acid of the general formula R.CO.COOH. The keto acid either acts directly with co-enzyme A to give the acetyl derivative or else by gradual oxidations successive units are broken off. In either case the product acetyl-co A is fed into the Krebs cycle and energy is made available.

FACTORS AFFECTING THE RATE AND PRODUCTS OF RESPIRATION

11. Temperature. As we have seen respiration is a series of enzyme-controlled reactions from certain of which energy is derived and made available to the cell. Enzymes are temperature dependent (*see* IV, **11**) and in general the rate of enzymic reactions doubles for an increase in temperature of 10°C (this is expressed as $Q_{10}=2$) until the point of denaturisation of the enzyme occurs.

For warm-blooded animals the optimum temperature of metabolic activity, including respiration, is approximately 37°C. Other poikilothermic animals and plants may have different optima: thus arctic algae respire efficiently near 0°C and those in hot springs nearer 70°C.

12. The respiratory substrate. The type of substrate being oxidised by a particular organism at any one time has an effect on the proportions of gases consumed and produced.

The ratio of carbon dioxide liberated to that of oxygen taken up in unit time is termed the *respiratory quotient* or *RQ* and a quantitative determination of this allows us to tell what the organism is respiring at any time. For carbohydrates (*see* IV, **3**) the RQ is $\dfrac{6CO_2}{6O_2} = 1$ while for a fat such as tri-olein,

$$C_{57}H_{104}O_6 + 80O_2 \rightarrow 57CO_2 + 52H_2O,$$

it is $\dfrac{57CO_2}{80O_2} = 0.71$. For proteins the RQ is approximately 0.8. Thus for an animal or plant using all three substrates RQs would vary between 0.7 and 1.0. For man the usual figure is about 0.85.

Anaerobic respiration producing CO_2 for no consumption of O_2 yields a theoretical RQ of infinity.

13. The availability of oxygen. The concentration of oxygen available determines which chemical pathway the pyruvic acid produced at the end of anaerobic glycolysis will follow. As a general rule above concentrations of two per cent it passes via acetyl-co A into the Krebs cycle while below this concentration it is converted to alcohol and carbon dioxide.

Many plant seeds, e.g. rice, enter anaerobic respiration for a period at the start of germination when their metabolic needs exceed the supply of oxygen that is available.

SPECIAL FEATURES OF RESPIRATION IN PLANTS

14. Energy demands in plants. Unlike animals the major part of many plants, especially woody ones, is not alive but is made up of dead supporting, conducting and protective tissues. These tissues do not respire. The different living tissues of plants may respire at very different rates from the high energy demands of growing regions to the minute quantities required by seeds and spores (*see* Table XII).

TABLE XII. OXYGEN CONSUMPTION IN A PLANT

Tissue	$mm^3 O_2$ per gram tissue wet weight per hour
Seed: dormant	0·00001
germinating	150
Leaf	332

Generally speaking, *the respiration rates of plants are substantially lower than those of animals* because their demand for energy is lower. Plants do not move (except for some lower plants and their gametes) nor do they conduct nerve impulses or have contracting cells, neither, on the whole, do they generate heat. The main energy consumption of the plant is for the synthesis of organic molecules, membranes and transport of substances.

Because of their low energy needs plants are better able to withstand temporary anaerobic conditions than animals.

15. Gaseous exchanges in plants. Small plants, such as many algae, have large surface areas relative to their volumes. They also tend to be aquatic, or, like fungi, to flourish only in damp places. For these plants the diffusion in and out of oxygen and carbon dioxide over their surfaces is sufficient to meet their respiratory needs.

Once inside the plant, whether by diffusion from the atmosphere or by generation in situ, oxygen can diffuse via intracellular spaces and be carried in solution in xylem and phloem.

In higher plants the outer layers are covered either with an impermeable epidermis or bark. Gaseous exchanges in such plants take place by means of *stomata* in the non-woody parts and by *lenticels* in the bark (*see* Fig. 76).

(*a*) *Stomata* (*see also* VI, **22**) are found on the leaves and the green parts of plants. There may be thousands to the cm². The stomata is itself a hole surrounded by two guard cells whose volume changes and by which the size of the hole is regulated.

When the plant is not photosynthesising, e.g. in the dark, the stomata will usually be closed but even so they allow small quantities of gases to pass in and out. When the stomata are open oxygen and carbon dioxide pass readily along their respective diffusion gradients. In normal conditions much of the oxygen required by the plant for its respiration will be generated by the photosynthesis of the plant itself.

Many freshwater plants have stomata connected to special aerenchyma tissue which allows diffusion of oxygen down to the roots of the plant which will be living in conditions of low oxygen pressures.

(*b*) *Lenticels* are interruptions in the bark where the suberised and impermeable cells are loosely packed and where it is possible for oxygen to penetrate to the deeper tissues. The intake of oxygen via the lenticels is necessarily slow, so too is its utilisation.

16. The relationship between respiration and photosynthesis. When the green plant is in the light, photosynthesis normally exceeds respiration by such a large amount that no escape of respiratory CO_2 can be detected. The CO_2 produced by respiration is fixed again by photosynthesis. The plant is, however, respiring all the time and a proportion of the phosphoglyceric acid, PGA (*see* VI, **26**) formed in photosynthesis goes at once into the aerobic phase of respiration yielding ATP for the cell.

In order to demonstrate that a plant is respiring its gas exchanges in the dark must be examined.

NOTE: Young seedlings show respiratory exchanges for some time before they start to photosynthesise.

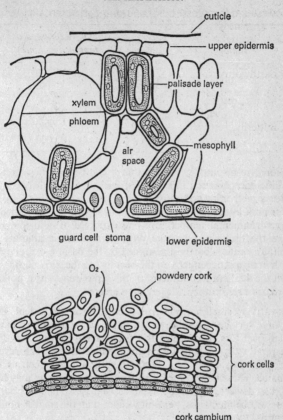

FIG. 76 *Gaseous exchanges in plants take place through the stomata of the leaves and green stems and via the lenticels of corky regions.*

Direct diffusion of oxygen into the young root tissues also occurs.

SPECIAL FEATURES OF RESPIRATION IN ANIMALS

17. Energy demands by animals. The oxygen consumption of a number of animals is in Table XIII. Oxygen consumption is pro-

TABLE XIII.   OXYGEN CONSUMPTION IN SELECTED ANIMALS

Animal	O_2 (expressed as $mm^3/g/per\ hour$)
Earthworm	60
Frog	150
Mouse, resting	2,500
active	20,000
Butterfly, resting	600
active	100,000

portional to respiratory rate so that it can be seen how great is the difference between individual species as well as between the same species resting and active.

For poikilotherms the respiration rate is largely determined by the environmental temperature and extremes of cold and heat define the limits of survival. With homiothermic animals the respiratory rate is partly determined by the basal metabolic rate which in turn is dependent on the heat lost to the surroundings.

On the whole a homiotherm of the same weight as a poikilotherm has a respiratory rate ten times greater.

NOTE: Very small homiotherms such as birds and rodents have a large surface area relative to volume and thus tend to have a high heat loss. Because of this they have to have a very high respiratory rate to maintain their temperatures. The reverse problem exists for large homiothermic animals such as elephants. These have some difficulty in losing heat and may have special devices, e.g. large ears, to assist losses. Their respiration rate will be much less, per unit weight, than the smaller homiotherms.

All homiothermic animals are capable of some degree of adaptation of their metabolic, and thus respiratory, rate according to the external temperatures. This is a long term adjustment of temperature control (see also IX, 12).

GASEOUS EXCHANGE IN ANIMALS

18. Respiratory surfaces. There are certain features associated with a respiratory surface which all animals have in common, and they are as follows:

(a) *The surface is large*, often by being much folded so as to present maximum contact with the respiratory medium, i.e. air or water.

(b) It is also *thin to speed diffusion*.

(c) *The surface is moist* so that oxygen and, to a lesser extent, carbon dioxide may pass through in solution.

(d) *The surface is often ventilated* so that fresh supplies of the oxygen-containing medium may be brought to it and stale air or water removed.

In larger animals which transport respiratory gases in their blood streams *the respiratory surface has a good circulatory supply*.

19. Small animals. The smaller animals such as protozoa, or those which have small volumes of living tissue, such as coelenterates and platyhelminthes, have sufficiently large surface areas relative to their volumes to allow simple diffusion to satisfy their O_2 needs. In such organisms the gradient of oxygen from the water to the tissues where it is being utilised is sufficient to allow a steady diffusion into the organism. Carbon dioxide formed as a result of respiration will diffuse out to the medium.

NOTE: The annelids have too great a bulk for this simple diffusion throughout the animal to suffice and they have a blood stream to carry the respiratory gases. Oxygen diffuses in through the thin cuticle and across the moist epidermis to the loops of surface capillaries that it contains. Oxygenated blood passes forwards in the dorsal vessel and is distributed by the ventral vessel to the deeper tissues.

If annelids such as earthworms dry up the permeability of their epidermis is reduced and they suffocate: on the other hand they can survive indefinitely submerged in well aerated water. Under normal conditions earthworms do not use the haemoglobin of their blood to carry oxygen which is merely dissolved in the liquid fraction.

In earthworms CO_2 is partly returned to the air via the blood and epidermis but a considerable amount is combined with Ca^{++} to form insoluble calcium carbonate in the calcareous glands. The lumps of $CaCO_3$ are passed into the gut.

20. Tracheal systems of insects. Insects are the most active of all animals and their ability to develop very high metabolic rates depends partly on the efficiency of their gaseous-exchange system.

(*a*) *The trachea.* The system consists of fine tubes, the trachea, which lead from the surface of the insect directly to its internal tissues. The trachea start at the spiracles which are provided with occlusor muscles allowing them to be opened or closed.

(*b*) *The tracheoles.* Near the surface the tracheal tube will be some 2μm or so in diameter but as it penetrates deeper it begins to branch into many finer tubes, 0·6–0·2μm in diameter, called tracheoles. These tracheoles are some 0·5 mm in length and terminate within the tissues. They do not have the epicuticle present and are therefore permeable to gases and liquids (*see* Fig. 77).

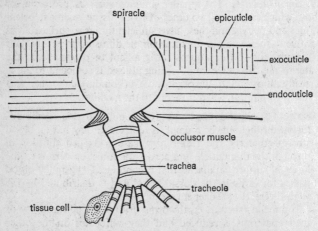

FIG. 77 *Insect spiracles lead into the tracheal system which break up into fine branches leading directly to the tissues.*

NOTE: When the individual insect is active respiratory carbon dioxide stimulates the opening of the spiracles and at the same time the accumulation of respiratory products, e.g. lactic acid, within the cells increases their osmotic pressure. The fluid that normally occupies the ends of the tracheoles is thus drawn into the cells so that the effective surface area for gaseous exchange is much increased.

(*c*) *Efficiency of diffusion.* The whole system depends on diffusion for its continued functioning and despite its very high efficiency over a short distance (i.e. not exceeding 2–3 cm) this is one of the factors that limits the size of insects (*see also* II, 6). The passive nature of a static tube system is improved in many insects by having some form of ventilation. This may be by movements of the body as in bees, or by having a type of "draught" passing from one end of the tracheal system to the other, e.g. dragonflies.

While all the insects' oxygen enters through the tracheal system only a part of the excretory carbon dioxide is lost in this way. Much of the latter is combined with nitrogenous wastes to make uric acid which is passed out from the Malpighian tubules into the gut.

GILLS

21. Definition. Gills are the respiratory surfaces of a number of aquatic animals including the chordates which all at one time or another had gills or gill slits during their development. Gills are found in lamellibranch molluscs among others, and also in many crustacea.

22. Fish gills and respiration. These, as seen in the dogfish or herring, consist of numerous flattened plates called filaments (*see* Fig. 78). Afferent vessels from the ventral aorta bring de-oxygenated blood and this is driven across the capillary beds of each filament. At the same time a current of water is maintained across the gill either by the raising of the floor of the branchial cavity or by the suction of the parabranchial (or opercular in teleosts) cavities. A continuous flow of oxygen-containing water is thus passed across the gill filaments. Haemoglobin in the red corpuscles of the blood combines with some of the oxygen that has diffused across the thin capillary, and the afferent vessels lead the oxygenated blood into a dorsal vessel whence it is distributed around the body.

Gills, like various other biological exchange systems, employ the principle of *counter currents*. The water passes in the reverse direction to the blood and thus high oxygenated water comes into contact with almost saturated blood and some more oxygen diffuses into the latter. If the flow had been parallel much less efficiency would be obtained.

The gills of the dogfish can remove more than half the oxygen content of the water passing across them and while this is more

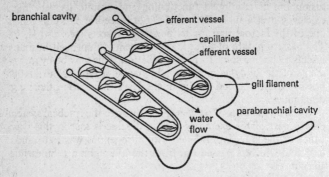

FIG. 78 *Diagram of a single filament of a dogfish gill showing the relationship of the capillary network to the raised lamellae.*

Fish gills may extract as much as eighty per cent of the oxygen from the water passing over them.

efficient than the mammalian lung it must be remembered that the oxygen content of water is at most one per cent compared with twenty per cent in air. There is a definite limit imposed on the metabolic rate of all animals that obtain their oxygen supply from water.

23. Tadpole gills. The newly-hatched tadpole develops external gills after some seven days and provides these with arterial shunts from the third, fourth and fifth arches. The external gills are only temporary projections of the skin. After the third week the internal gills develop with a blood supply from the third to sixth arches. These gills operate in the same way as the gills of fishes, and, like the teleosts, they are protected by an opercular cover. During the metamorphosis from the tenth to the twelfth weeks the internal gills are replaced by lungs.

24. Gills in other animals. Certain lower chordates such as *Amphioxus* have gills which are a series of slits in the pharynx. The supporting arches between the individual slits are well supplied with blood and the gill provides a large respiratory surface. Ciliary currents on the gill arches also provide a feeding current for the capture of small particles which are wafted down the alimentary canal.

This combination of a respiratory and feeding surface is also a common feature of lamellibranch molluscs such as *Anadon*, the freshwater mussel, and *Mytilus*, the mussel. Oxygen from the incoming current of water diffuses across the large gill surface while at the same time particles from the water are extracted by ciliary currents and mucus and are led to the mouth.

Higher Crustacea, e.g. the crayfish, lobster, crab, etc., have well-developed gills which are filamentous outgrowths from certain of the segmental appendages, more particularly those of the claws and first three walking legs. These outgrowths are protected under the chitinous carapace and a special appendage, the maxilla, is instrumental in creating a current which flows over the gills. As with other gill systems an afferent and efferent blood supply is present and a respiratory pigment called haemocyanin has a role equivalent to the haemoglobin of vertebrates.

25. General characteristics of gill systems. From the above examples it can be seen that gill systems tend to have certain common features. They are developed in aquatic animals because the water allows them support, and they tend to be evaginations of the body surface rather than invaginations like lungs.

Some form of protection such as a carapace or operculum is present as gills are delicate structures. A means of ventilation is found so that the respiratory medium may continuously be renewed and waste and de-oxygenated water removed.

The gill is richly supplied with blood containing a respiratory pigment which passes through fine capillary networks across which gaseous exchanges can take place.

LUNGS

26. Mammalian lungs. The lung of a mammal such as man has a very large surface area, being made up of some seven hundred million minute air sacs or alveoli. Each of these is approximately 0·1 mm in diameter and has a thin wall 0·5μm in thickness. The walls of the alveoli are supplied by capillaries from the pulmonary artery and are kept moist be secretion. Oxygen passes across from the air sac into the blood vessels and thus to the red corpuscles. At the same time water and carbon dioxide diffuse in the reverse direction. Lung tissue is rich in carbonic anhydrase which facilitates the breakdown of carbonic acid into carbon dioxide and water.

Ventilation of the human lung is affected by the raising of the rib cage and the lowering of the diaphragm muscle. This results in a reduction of the pressure in the pleural cavity from —0·5 kPa to —1·0 kPa. The reduction of pressure is communicated to the lung and air is drawn in. As it passes through the nasal passages, trachea and bronchi it is warmed and filtered.

Normal volume exchange in man is some 500 cc and ventilation takes place sixteen times per minute. Under exertion, as much as 3,000 cc may be exchanged at each breath and the ventilation rate can also be increased. O_2 passes continuously across the alveolar walls.

Co-ordination of ventilation volume and rate is centred in the medulla of the brain.

27. Lungs in other animals. Lungs are found in land-living vertebrates. They are invaginations of the body surface. Being inside the animal they can be supported, kept moist and protected from mechanical damage.

Within the vertebrates lungs first appeared in the lung fishes and related forms of the Devonian period. They are formed by a modification of the back of the pharynx and are homologous with the swim-bladder of teleosts. Vertebrate lungs are supplied with blood by the sixth arterial arch, the pulmonary.

Frogs have lungs whose capacities (compared with the animal's volume) is only $\frac{1}{16}$ that of mammalian lungs. The air is forced down into them by a buccal force pump and the lung is only used at certain times of the year. Gaseous exchanges take place all over the body of the frong, because the skin is permeable.

Reptiles have a more efficient form of lung with a vacuum intake created by rib movements as in mammals. They do not possess a diaphragm. Birds on the other hand have tubular lungs with a very efficient continuous-flow system of ventilation. They are able to supply oxygen to the blood at a greater rate than the lung of the mammal and birds are remarkable for their very high metabolic rates. Such rates are essential to provide energy for flight, the most demanding of all forms of locomotion.

A few other animals, such as the pulmonate gastropods, also have lungs. These are in the form of cavities in the mantle whose walls are well supplied with blood vessels. There is little active ventilation and respiratory rates of such animals are low.

TRANSPORT OF RESPIRATORY GASES IN ANIMALS

28. Oxygen. The high respiratory rate of many animals can only be met by the rapid transport of large amounts of oxygen from the respiratory surface to the metabolising tissue. In such animals there is a respiratory carrier in the blood because of the low solubility of oxygen in water, or plasma. Two important carriers are the following:

(*a*) *Haemocyanin.* This is a colourless or blue copper-containing substance which is capable of accepting and transporting oxygen by conversion of part of the cuprous to the cupric state. It has the same shape of dissociation curve as haemoglobin (*see* (*b*)) but is not able to transport equivalent volumes of oxygen. It is found in many crustaceans and some molluscs such as the cephalopods (squids, etc.). It is not confined to corpuscles but exists freely in the blood plasma.

(*b*) *Haemoglobin.* This has a very wide distribution and is found in many phyla including the molluscs, annelids, arthropods and chordates. It is even found in some protozoans. The haemoglobin is an iron-containing pigment, the haem part, attached to a protein, globin, giving a molecule of some 70,000 m.w. The iron exists in the ferrous state whether the pigment is oxygenated or not, so the actual method of oxygen transport is complex.

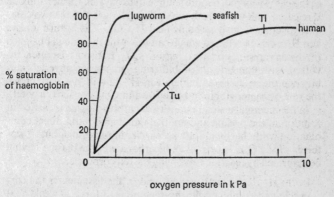

FIG. 79 *Oxygen dissociation curve for some haemoglobins.*
Tu = tension unloading, Tl = tension loading. 10 kPa is approximately equal to 80 mm Hg.

One of the important properties of haemoglobin which allows it to act as a respiratory carrier is that the amount of oxygen with which it combines depends on the partial pressure of oxygen available. At maximum oxygenation human haemoglobin will combine with up to four O_2 molecules which it does at a partial pressure of oxygen of approximately 9·3 kPa (70 mm Hg). The reations are readily reversible, i.e.,

$$Hb + O_2 \rightleftharpoons HbO_2 + O_2 \rightleftharpoons HbO_4 + O_2 \rightleftharpoons HbO_6 + O_2 \rightleftharpoons HbO_8$$

Other haemoglobins combine and dissociate with oxygen at very different partial pressures and to be of use to the organism it is necessary that the haemoglobin fully combines with the oxygen at that partial pressure at which it is available in the environment. Similarly it must dissociate at the partial pressure of oxygen existing in the animal's tissues.

This can be seen in Fig. 79. It should be noted that although the haemoglobin is shown as one hundred per cent saturated in each case, it does not mean that it is actually carrying the same amount of oxygen. Thus while fully saturated human blood can carry as much as twenty-six volumes oxygen per hundred volumes of blood, worm blood will carry less than five.

29. Carbon-dioxide transport. The transport of carbon dioxide is, at least in theory, more likely to be possible by blood fluid alone as the gas is very much more soluble than oxygen. Carbon dioxide is released with water from the respiring tissues and the two combine together to give carbonic acid H_2CO_3. This in turn ionises and H^+ ions are released which can upset the delicate pH balance of the organism. For this reason CO_2 transport is often involved with some form of buffering. In marine invertebrates blood proteins assist as buffers for H^+ but in the vertebrates the role of haemoglobin in buffering and CO_2 transport is vital.

In the mammal a small quantity of carbon dioxide is carried in solution in the plasma and another small amount in direct combination with haemoglobin as carbamino-haemoglobin. Two-thirds of the CO_2 is carried as bicarbonate in the plasma having been formed by reactions involving the red corpuscle.

NOTE: H^+ ions from carbonic acid in the tissue enter the corpuscle and combine with haemoglobin to form a weak acid which we may call HHb. These H^+ ions are thus removed from solution while at the same time HCO'_3 ions from the dissociated carbonic acid are released from the red corpuscle to the plasma. This release of negatively charged ions from the corpuscle is accompanied by the influx of Cl' ions from the plasma.

On arrival at the lungs the bicarbonate re-enters the red corpuscles and once again combines with H from the dissociation of the HHb forming carbonic acid. This is broken down in the presence of carbonic anhydrase to give carbon dioxide and water which are liberated into the alveoli and then to the atmosphere. Cl' ions leave the corpuscle.

NOTE: Oxygen and carbon-dioxide transport are closely linked in vertebrates as the presence of carbon dioxide causes a decrease in the amount of oxygen that can be carried at any given partial pressure. The effect of this is to stimulate the release of oxygen in the tissues where the CO_2 is high and the loading of oxygen at the respiratory surface where it is low.

THE MEASUREMENT OF RESPIRATION RATES

30. Weight change methods. As respiration may be represented by the simple equation

$$C_6H_{12}O_6 + 6O_2 \rightarrow 6H_2O + 6CO_2$$

it is possible, in theory, to estimate its rate by measuring the change in any of the reactants or products. In practice some are easier to measure than others.

Loss of carbohydrate can be measured in plants by taking a portion from a leaf and drying it to a constant weight, A. After a given period in the dark an exactly similar portion is taken and treated in the same way. This will have weight B and the difference in A and B is mostly due to respiration loss. This method does not take into account translocation into and out of the leaf and it is difficult to perform accurately.

31. Gas volume changes. The CO_2 produced in a given time may be estimated by collecting the air that has passed over the organism and bubbling it through barium hydroxide (plant tissues are to be kept in the dark). The $Ba(OH)_2$ combines with the CO_2 in the outgoing air to form barium carbonate and the quantity of this can be determined by titration.

Another method of measuring CO_2 production is by volume change. The organism or respiring tissue is enclosed in a flask attached to a manometer. Within the flask is KOH which will take up any CO_2 expired. This is the basis of the Warburg and Barcroft respirometers and it is of very wide application.

Results are expressed as cc of CO_2 per g of dry weight tissue per hour or similar units as may be appropriate.

PROGRESS TEST 8

1. Respiration is not the same as breathing. Why not? **(1)**
2. What is the importance of ATP? **(2)**
3. What changes to the glucose molecule take place during anaerobic glycolysis? **(4)**
4. What is the importance of the Krebs cycle? **(5)**
5. Why is cytochrome essential to aerobic respiration? **(5)**
6. What role does the mitochondrion play in respiration? **(6)**
7. Why is anaerobic respiration inefficient? **(7)**
8. How are fats fed into the respiratory pathways? **(9)**
9. How does a 10°C rise in temperature affect the rate of respiration? **(11)**
10. What is the importance of lenticels to woody plants? **(15)**
11. What are the important properties of respiratory surfaces? **(18)**
12. Why does a tracheal system impose limitations on size? **(20)**
13. How do fishes maintain a current of water across their gills? **(22)**
14. How many alveoli are in the human lung? **(26)**
15. Why do birds need such efficient lungs? **(27)**
16. What is haemocyanin? **(28)**
17. Draw the loading curve for haemoglobin. **(28)**
18. How is the bulk of carbon dioxide carried in the blood? **(29)**
19. How may the respiratory rate be measured? **(30, 31)**

EXAMINATION QUESTIONS

1. Describe the structure and function of mitochondria and chloroplasts (*see also* Chapter IV).

London, 1975

2. Study the following information which summarises the metabolic roles in the mammalian body of certain vitamins.

Vitamin A combines with the protein opsin to form visual

purple in the retina, is important in the normal functioning of epithelial cells, and is involved in the synthesis of certain cell proteins.

Vitamin B1 gives rise to the enzyme which is necessary for the oxidation of pyruvic acid.

Vitamin C is involved in the manufacture of the intercellular cement which holds cells together.

Vitamin D is involved in the ileum where it unmasks a region of the epithelial cell deoxyribonucleic acid which is responsible for the synthesis of a protein which transports calcium and phosphorus.

From this information and from your knowledge of physiology attempt to describe the symptoms which might be expected to result from a deficiency of each of these vitamins.

For each vitamin, proceed as follows:

(*i*) name it (e.g. vitamin A),

(*ii*) describe the symptoms likely to arise as a result of it being deficient,

(*iii*) briefly explain the metabolic reasons for the occurrence of these symptoms.

Associated Examining Board, 1976

3. "Respiration is vital to all living organisms because it provides the energy required not only to maintain their body structure but also to carry out their varied activities. The energy is liberated by oxidative processes in the cells and for this purpose the oxygen is taken in at a respiratory surface where carbon dioxide is usually liberated. It is convenient, therefore, to separate *external* from *tissue* or cellular respiration, but in animals the size of vertebrates a third *transport* process is required." (*Marshall and Hughes*, 1965.)

Explain this passage in language which could be understood by an intelligent person who has never been taught any biology, chemistry or physics. *Do not* give a full account of the chemistry of respiration.

Associated Examining Board, 1974

4. Some properties of water and air are given in the table on p. 188.

What problems do the above data pose when terrestrial and aquatic animals ventilate their respiratory surfaces? Illustrate how these problems have been overcome by referring to the mechanisms of ventilation used by a variety of *named* organisms.

Associated Examining Board, 1977

	Water	*Air*
Oxygen content	about 1%	20%
Oxygen diffusion rate	low	high
Viscosity	×100 that of air	low
Density	Specific gravity ×1000 that of air	very low

5. In both photosynthesis (Calvin cycle and cyclic photophosphorylation) and in respiration (Krebs cycle) key biochemical pathways are based on cyclic systems such as

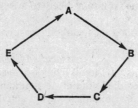

rather than sequence chains such as A → B → C → D → E.

Briefly review the major cyclic pathways in the important physiological processes mentioned above and attempt to deduce why a cyclic biochemical system might be preferable to a chain one. You are *not* expected to show detailed knowledge of biochemical formulae.

Associated Examining Board, 1975

6. Discuss the differences between aerobic and anaerobic respiration in living organisms.

after Oxford and Cambridge, 1977

CHAPTER IX

Important Maintenance Activities

INTRODUCTION

1. The concept of homeostasis. This is the term that is used for the *maintenance of the steady state* in the animal body (*see* Fig. 80). Plants may also be considered to show homeostatic adjustment to such stimuli as light, *see* XI, 37, but owing to their sessile nature their systems of physiological co-ordination are mainly much less elaborate than those of animals.

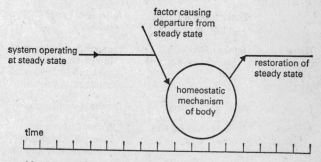

FIG. 80 *A scheme to show the nature of a homeostatic adjustment from departure from the steady state.*

Such systems are used widely in the maintenance of a constant internal medium by all living things but are particularly highly developed in mammals.

Homeostasis is the means whereby physiological systems operate both separately and together to buffer against fluctuations from the optimum conditions.

There are countless examples of homeostatic control systems in all animals. Some of the important instances of homeostasis in the mammal are given in Table XIV.

The above adjustments are all co-ordinated by endocrines or the autonomic nervous system from control centres in the brain.

TABLE XIV.   HOMEOSTATIC CONTROL SYSTEMS

System	Cause of fluctuation	Homeostatic control
Circulation	Muscular activity	Increase of heart beat, blood pressure and supply of blood to the muscles concerned.
	Temperature change	Adjustment of blood supply to or from the skin. Sweating or shivering.
	pH changes due to metabolic activities	Buffering by haemoglobin, phosphates, etc. in blood. Release of H^+ or OH' by kidney.
	Sugar taken up from gut or by activity	Adrenalin–insulin balance.
Kidney	Intake of variable quantity of water or ions	Composition and strength of urine controlled by mineralocorticoids and antidiuretic hormone.
Lung	Muscular activity	Increased ventilation.
Muscles	Balance, posture and locomotion	Changes monitored by main sense organs or proprioceptors lead to feedback and appropriate adjustments in antagonistic systems. Control centre is cerebellum.
Digestion	Presence of food	Secretion of enzymes, peristaltic activity, and increased blood supply.

In a broader sense homeostasis may also be said to apply to situations involving the whole organism such as the response of the body to pain, danger and stress.

Mammals are highly evolved animals and their bodies function within very narrow physiological limits. If homeostatic control breaks down death rapidly follows. Thus raising the temperature of the human body by only 3°C causes widespread functional disorganisation. The ability of mammals to maintain their bodies at a steady state despite widespread fluctuations in the environment is one of the major factors determining their success.

In the present chapter certain of the more important maintenance activities of mammals and other animals which lead to a preservation of a constant internal medium will be described. The activities selected are as follows:

(a) The circulation of the blood in mammals.

(b) The regulation of temperature.

(c) Excretion: the maintenance of a constant osmotic and ionic internal medium and the elimination of waste.

THE CIRCULATION OF THE BLOOD IN MAMMALS

2. The heart. The mammalian heart is a four-chamber pump (*see* Fig. 81). It is separated into two halves in such a way that a double circulation takes place and that oxygenated and deoxygenated blood do not mix.

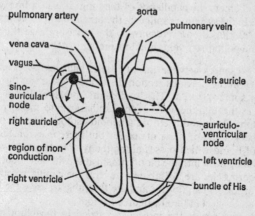

FIG. 81 *A simplified diagram of the mammalian heart to show the means of co-ordination of beat.*

The homology of vertebrate hearts is shown in Fig. 152.

Birds also have a double circulation unlike any of the other classes of vertebrates. Both mammals and birds have very high metabolic rates which require complete separation of the two types of blood so that an efficient oxygen supply to the tissues is ensured.

(a) *The heart beat* is initiated by the *sino-auricular node* in the right auricle. This pace-maker generates impulses automatically but may be slowed down by the vagus depressor (parasympathetic) or speeded up by the adrenalergic (sympathetic) system. A wave of contraction passes across the auricles and sets up excitation in the auriculo ventricular node. From the latter impulses pass down the auriculo ventricular bundle to the bases of the ventricles causing them to contract.

(b) *The output of the heart* of man is some five litres per minute but this may be increased some eight times under exertion. The normal pulse rate is 72 per minute. Very small mammals have enormously high rates of heartbeat, up to 1,000 per minute, associated with their high metabolism.

(c) *Pressure in the heart* at systole is about 3·3 kPa (25 mm Hg) in the right ventricle and over 13·3 kPa (100 mm Hg) in the left. This pressure wave is transmitted to the arteries which have muscular walls and there it becomes smoothed out. By the time the blood enters the capillaries of the lungs it is at a pressure of 2·6 kPa. In the systemic capillaries the pressure will be at approximately 4·0 kPa. Blood returning to the heart both via the venae cavae and pulmonary veins is at low pressure of 0·67 kPa.

3. The exchanges between the capillaries and the tissues. As we have seen in the previous section the blood pressure in the capillaries is some 4·0 kPa. This hydrostatic pressure forces small molecules through the walls of the capillaries into the surrounding lymph. The blood itself has a colloid osmotic pressure which draws back some of these molecules but there is a considerable excess of liquid which passes along the lymphatic ducts eventually draining back into the blood stream at the thoracic duct.

Red corpuscles are not able to traverse the capillary walls but white corpuscles can do this and are thus able to leave the blood stream and collect at the site of an infection.

The lymph passes the molecules carried in the blood to the tissues and waste substances from the tissues diffuse back into the blood (*see* Fig. 82).

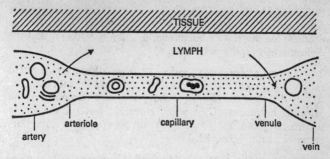

FIG. 82   *The relationship of the blood to the lymph and the surrounding tissues.*

At the arterial end of the capillary the hydrostatic pressure of the blood exceeds the osmotic pressure by 1 kPa (8 mm Hg). Plasma substances, water, ions etc. are thus squeezed out into the lymph. At the venous end of the capillary the osmotic pressure of the blood exceeds the hydrostatic pressure by 2·4 kPa (18 mm Hg). Substances of small molecular weight such as water, ions, urea etc. are thus drawn back into the blood.

THE FUNCTIONS OF THE BLOOD

4. Transport of oxygen and carbon dioxide. This has been covered (*see* VIII, 28). The haemoglobin is carried in the erythrocytes or red corpuscles which number some 5×10^6 mm³. Erythrocytes are biconcave discs approximately $7\mu m \times 2\mu m$. In mammals they have no nuclei and after a variable time in the bloodstream the red corpuscles break down and some of the iron of their haemoglobin is recovered. A breakdown product of the latter is bilirubin.

5. Transport of food. Monosaccharide sugars, amino-acids and fats in one form or another are absorbed into the capillaries of the ileum. These unite in the hepatic portal vein which carries the assimilated food to the liver. According to the demands of the body food substances will be released into the hepatic vein and enter the general circulation. The hormones adrenalin, insulin and the glucocorticoids control the concentrations of sugars in the blood. This does not exceed 0·1 g per 100 ml.

Besides food substances the blood carries the ions K^+, Na^+, Cl', Mg^{++}, Ca^{++}, SO_4'' PO_4''' and HCO_3'

6. Transport of waste. Besides bicarbonate from tissue respiration the blood carries small amounts of nitrogenous waste products from the various organ systems to the kidneys. These include urea, 30 mg per 100 ml, ammonia, 0·02 mg per 100 ml, and the endogenously produced creatinine.

7. Transport of hormones. Endocrine organs have a rich blood supply and secrete their hormones directly into the bloodstream. At any one time the blood will carry very small amounts of the principal hormones of the body. Hormone levels feed back to the master gland, the pituitary, which adjusts the trophic hormones accordingly (see XI, 17).

8. Transport of heat. The blood allows the transfer of heat from the deeper tissues to the surface of the body where it can be lost.

9. Defence against infection. Infections of the body are mainly due to the invasion by bacteria and viruses and by the toxins these organisms produce. Both the organism and the toxin are classed as antigens and the body reacts against these by the formation of antibodies.

There is a lot of new work in this area and it is not easy to summarise it in the brief space available. Some of the essential ideas however are as follows:

(a) Antigens enter the body and are picked up at the surface of special defence cells called macrophages. Here T lymphocyte cells (mainly derived from the thymus) are able to "recognise" the particular chemical nature of the invading antigen.

(b) The T cells now switch on genes in certain bone marrow derived cells (called B lymphocyte cells) which are thus caused to multiply and produce a specific antibody to match the antigen.

(c) Antigen and antibody combine to form a harmless complex and this becomes phagocytised by blood and lymph phagocytic white corpuscles. In the blood the polymorphs and monocytes (see Fig. 83) are specialised as phagocytes, while there are also lining cells in the liver and spleen which do the same job.

(d) The setting up of the T cells is clearly a most significant part of the whole process of combating antigens. This is done in the young life of the mammal where it learns to distinguish between its own protein systems, against which it does not react, and foreign proteins, against which it does.

10. Antigen–Antibody reactions. Antibodies belong to a class of protein called immunoglobulins. Recent investigations show

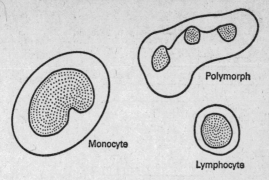

FIG. 83   *The cellular constituents of blood.*

Monocytes and polymorphs are white blood corpuscles which are actively phagocytic, while lymphocytes (which may be large or small) are made in the lymphatic system whence they may enter the blood stream. They are secretory. Platelets are involved in clotting.

these proteins consist of a series of four parallel chains of amino-acids linked by sulphur cross-bridges. The two central chains are longer than the outside ones and a good deal of the composition of all four chains is fixed.

At the end of the immunoglobulin that binds with the antigen all four chains have a series of variable sequences which can be modified to fit exactly on to a particular antigen (*see* Fig. 84). It is estimated that the human body can produce as many as 100,000 variations in antibody structure.

11. Note on immunity. There are various kinds of immunity and they are as follows (*see* Fig. 85):

(*a*) *Hereditary immunity.* This is built in to the genetic code, and can apply to a species, a race or an individual. It implies an ability to react against a substance, having had no previous contact with it. Examples of hereditary immunity are as follows.

(*i*) *Species.* Cats are naturally unaffected by the diphtheria bacterium, *Corynebacterium diphtheria.*

(*ii*) *Race.* A European shows a different degree of severity

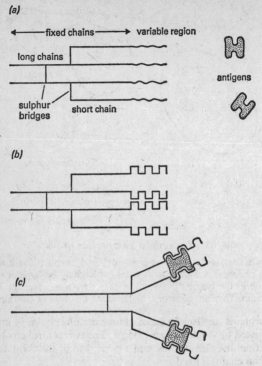

FIG. 84   *Antigen–antibody reactions.*

(*a*) Basic antibody protein. (*b*) Antibody becomes tailored to attach to a particular antigen. (*c*) Antibody and antigen combine and form a neutral complex.

of tuberculosis to an African, even though they are both infected with the same bacteria, *Mycobacterium tuberculosis*.

(*iii*) *Individual.* An individual may show an innate resistance to a disease e.g. flu.

(*b*) *Naturally-acquired immunity.* This is where the body has built up an immunity to a disease through natural contact with it.

(*i*) *Actively-acquired natural immunity.* A man who has had measles is immune to further attacks, as his body contains a permanent pool of T cells and antibodies (*see* 9) acquired as a reaction to the first attack. Alternatively, repeated contact with

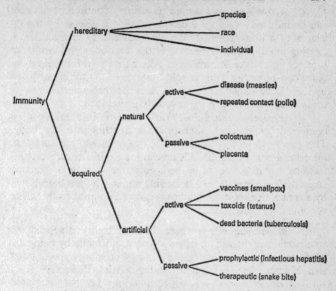

FIG. 85   *An outline of the various types of immunity.*

Examples of the various forms of immunity are shown down the right-hand side of the diagram.

a disease can set up this immune defence, even though the disease has not actually been contracted. For example, Victorian children were often immune to polio although they had never caught it. This was due to the fact that the disease was prevalent at the time and the children often came into contact with the virus. Many individuals also become immune to tuberculosis by a similar active process.

(ii) Passively-acquired natural immunity. The foetus is rendered immune to the same diseases as its mother as antibodies can pass across the placenta. After birth this immunity continues until the third or fourth month of life, after which time the infant's own immune system starts to take over. This process is complemented by a transfer of certain other antibodies in the colostrum (the first secretions of the mother's mammary glands).

(c) Artificially-acquired immunity. This is where substances are deliberately inoculated into the body, inducing or conferring immunity to an organism or range of organisms.

(*i*) *Actively-acquired natural immunity.* The immune system can be induced to set up a defence against a disease by the injection of a non-virulent or closely related form of the disease-causing organism, as is the case of the use of cowpox viruses to produce immunity to smallpox. Alternatively, the immune defence can be set up using toxoid (an inactivated toxin produced by the disease-causing organism), as is the case with tetanus, or using dead bacteria, as is the case with tuberculosis.

(*ii*) *Passively-acquired artificial immunity.* Immunity can be conferred by injecting appropriate antibodies into the body. This can be done as a prophylactic (preventative) measure, e.g. injecting gamma globulin as a precaution against contracting infectious hepatitis, or as a therapeutic measure, e.g. injecting snake venom serum from a horse into a man who has been bitten by a snake, the horse having been rendered immune to the snake venom.

(*a*) *Natural immunity* is where the body has become immune to a particular disease on its own accord, normally by being exposed to the pathogenic organism and by building up antibodies to it. Many people are immune to tuberculosis in this way.

(*b*) *Artificial immunity* is where immunity has been acquired by the body owing to human intervention. It may be *active* or *passive*.

(*i*) *Active immunity* is where a weakened or dead pathogen is introduced into the body. The defence systems build up antibodies as if the pathogen were active, and these antibodies remain in the bloodstream for a varying period. If the body is exposed to an active pathogen of the same type the circulating antibodies are able to destroy it. There is also an indication that the body "remembers" how to make a particular antibody once it has had occasion to make it. Examples of active immunisation are for typhoid, polio and cholera.

(*ii*) *Passive immunity* is where an antibody or antitoxin which has been synthesised in the body of another animal is injected into a human body. Thus if horses are given progressively increasing doses of snake venom they build up large quantities of snake antitoxin in their own blood. This can be extracted and injected into a man who has been bitten by a snake where it effectively neutralises the venom.

12. Clotting. When exposed to the air the blood clots which prevents loss of blood from a wound and also seals it against the invasion of micro-organisms. Clotting takes place as shown in Fig. 86.

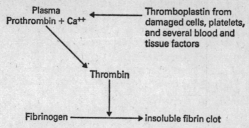

FIG. 86   *Processes involved in blood clotting.*

NOTE: The above scheme is much simplified, as there are a number of factors that are tripped off, one after another, for the eventual production of thrombin from prothrombin. All these factors need to be present for clotting to occur and haemophiliacs (whose blood does not clot readily) tend to lack factor VIII: thus prothrombin cannot be converted to thrombin. Presumably such a complex cascade mechanism helps to ensure that the blood does not clot by chance inside the body.

REGULATION OF TEMPERATURE

13. Temperature control and homeostasis. An important aspect of homeostasis in birds and mammals is the *maintenance of a constant internal temperature*. This allows the temperature-dependent enzymic reactions of the metabolism to proceed at their most efficient rates. A constant temperature also allows birds and mammals to live in a very wide range of geographical habitats.

Organisms that can control their temperature are called *homiothermic* while those that cannot are called *poikilothermic*. The relationship between internal and external temperatures in these two classes is illustrated in Fig. 87. It should be noted that homiothermy involves an increase in the basal metabolic rate by a factor of at least ten times and despite its great biological advantages it has to be "paid for" by a very substantial increase in food consumption and possibly by a shortening of the life-span as well.

14. New terminology. While the terms used in **13** are still per-

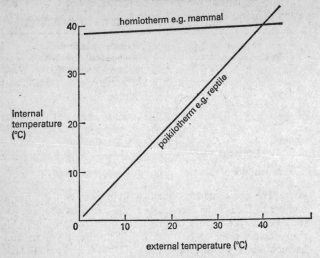

FIG. 87   *The relationship between internal and external tempera-ture in homiotherms (endotherms) and poikilotherms (ectotherms).*

missible some new ideas and definitions have entered this area of physiology. These are detailed in Table XV.

15. Relationship between body size and homiothermy (endothermy). Heat is lost from the body during breathing and excretion and by radiation from the whole surface. On the whole the primary problem of most homiotherms is the *conservation of heat* and we find that their bodies are covered with insulating fur (most mammals), feathers (most birds) or sub-cutaneous fat (whales, seals, penguins). Heat loss is increased by the evaporation of sweat from the skin or of saliva from the tongue.

The body surface of an animal increases as the square of its dimensions whereas the volume increases by the cube. This means that very small animals have proportionately a very large sur-face area and have formidable problems of heat conservation while very large animals have the reverse problem of encouraging heat loss.

(a) *Small homiotherms (endotherms).* Small birds and mammals have a very high metabolic rate in order to maintain their body temperatures. Thus a shrew may have a heart beat of over 1,000

TABLE XV.   TEMPERATURE REGULATION IN ANIMALS

Term	Definition and example
Endotherms	High metabolism and generation of heat coupled with good insulation. Mammals and birds are good examples.
Heterotherms	Like endotherms these produce heat and have good insulation but their temperature fluctuates, either during a long hibernation, as for example hedgehogs, or over a 24 hour period, as in bats.
Ectotherms	Animals that do not produce much internal heat and have little insulation. Their temperature will tend to fluctuate with their surroundings. Ectotherms may show behavioural adaptations in avoiding extremes of environmental temperatures.

per minute and needs to eat more than its own weight in twenty-four hours; the same figures apply to humming birds.

Some small homiotherms, now better termed heterotherms, such as bats and tropical birds that are very tiny, e.g. the weaver and humming bird, go into periods of temporary hibernation during the period in which they are not active and in temperate climates many small mammals, e.g. dormice and hedgehogs, go into hibernation.

(b) *Large homiotherms* (*endotherms*). Very large homiotherms such as the elephant have a great volume but a small surface area for heat loss. A variety of behavioural and physiological devices are used for heat loss such as flapping the richly vascularised ears.

16. Temperature control in extreme conditions. The most remarkable examples of homiothermy (endothermy) are seen in mammals living in those parts of the world where extremes of temperature are found, such as deserts and the land and sea of arctic regions. The basal metabolic rate of homiotherms (endotherms) is related to the environmental temperature and this allows adjustment of the heat production to the ambient temperature (*see* Fig. 88).

Mammals, such as whales, living in extreme cold have thick coats of blubber and a counter-current exchange to the flippers

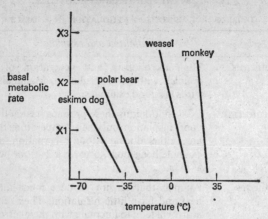

FIG. 88 *Temperature regulation by animals living in climatic extremes.*

so that returning blood is warmed to the temperature of the body before entering the main circulation.

In very hot environments such as deserts mammals such as camels, jack-rabbits and kangaroo rats have means of heat loss without the evaporation of large amounts of water. All these desert mammals have nephrons with long loops of Henle which are able to produce a very hypertonic urine.

17. Temperature control in man. Man loses approximately $1.2 KJ/cm^2/hr$ through his skin at 20°C. Thus he must make up at least this heat loss by his basal metabolism. Heat may also be drawn from storage when 'the body is cold. Heat will be lost by evaporation but may be gained or lost by radiation. However for constant temperature all these possible gains and losses must balance out.

(*a*) *General features.* Heat is retained in man by the subcutaneous insulation and the withdrawal of blood from the surface capillaries. The supply of blood to the latter can be decreased by a factor of one hundred times in cold conditions.

Heat is lost mainly by radiation and by the evaporation of sweat from the surface of the body. Under normal conditions only a litre or so of sweat is lost in a day but in extreme heat as much as four litres may be lost in an hour. Such a high rate of sweating causes the loss of 12 g of mineral salts in the same period. If more fluid is not taken in, the body rapidly becomes dehydrated and after ten per cent dehydration the blood becomes too viscous to transport heat sufficiently rapidly. An explosive heat rise therefore takes place and death results.

(b) *Co-ordination of temperature control.* Peripheral temperature receptors in the skin pass information to the central nervous system about environmental temperatures. More important in bringing about temperature changes in the body is the blood circulating near the hypothalamus of the brain. The temperature of this blood is monitored and the hypothalamus is the control centre for the initiation of appropriate activity leading to temperature change.

EXCRETION

The maintenance of a constant ionic medium and elimination of wastes are also vital homeostatic functions.

18. Definition of excretion. In its strict sense the term excretion is reserved for the *elimination by the organism of waste products of its metabolism.* Thus any waste product that has resulted from chemical reactions within the organism, e.g. CO_2, is excretory whereas the bulk of waste from the alimentary canal is not an excretory product as it has not been produced by metabolism.

NOTE: While such a definition allows recognition of excretory products there are a number of compounds which fit in several categories. Thus insect chitin is secretory as a structural material but contains excretory nitrogen compounds lost at moulting. The oxygen released by plant photosynthesis is a true excretory product and yet may immediately become utilised in aerobic respiration. Metabolic water is excretory and yet may be the only source of water available to certain desert animals, e.g. kangaroo rat.

On the whole the excretory products of organisms have either a depressing effect on their metabolism, e.g. CO_2 on plant respiration, or else are definitely toxic, e.g. alcohol and nitrogenous excretions.

19. Respiratory waste products. As a result of downgrading reactions, the energy content of the various possible respiratory substrates, carbohydrate, fat, protein, alcohol, etc., is transferred to stable energy carriers in the cell. Some energy is released as heat. The result of these downgrading reactions is the production of low-energy waste products such as carbon dioxide and water, the former from citric acid cycle and the latter from the combination of hydrogen and oxygen across the cytochromes.

In anaerobic respiration larger molecules are produced as metabolic wastes. These include pyruvic and lactic acid and various aldehydes and alcohols.

20. Nitrogen-containing waste products.

(a) *Nitrogen waste.* The excretion of nitrogen-containing waste products is typical of animals and these are derived partly from the deamination of excess amino-acids taken in with the diet (exogenous source) and partly from the breakdown of the animals' own proteins and nucleic acids (endogenous source).

Plants may also produce nitrogenous wastes such as ammonia but as they can synthesise the amino-acids they require, an excess of these is not normally made. Where deamination does take place the plant can store the ammonia produced in the form of amides and use the nitrogen as required. Deamination of proteins can be represented as follows:

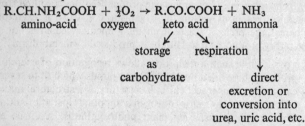

Which nitrogen containing compound is excreted depends largely on the animals' environment.

(b) *Specific nitrogen wastes.*

(i) *Ammonia* NH_3 is very soluble but also very toxic and can only be excreted in very dilute solution. It is the major nitrogenous excretion of many aquatic animals such as crustaceans.

(ii) *Urea* $CO(NH_2)_2$ is a common excretory product found in aquatic animals, e.g. fishes, and those terrestrial forms such as mammals which are not well adapted to water conservation.

NOTE: The conversion of ammonia into urea takes place in the liver or equivalent organ of the animal and may be represented

$$2NH_3 + CO_2 \rightarrow CO(NH_2)_2 + H_2O$$

The stages in this reaction involve the "Ornithine cycle" (*see* Fig. 89).

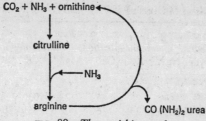

FIG. 89 *The ornithine cycle.*

(*iii*) *Uric acid* and its salts, the urates, are very insoluble and much less toxic than either ammonia or urea. They can therefore be excreted without loss of water and are the major nitrogen products of those groups which conserve water as one of the ways of survival on land. Reptiles, birds and insects all use this form of eliminating nitrogenous waste. It is not very clear how uric acid is derived from ammonia but its formula may be represented as in Fig. 90.

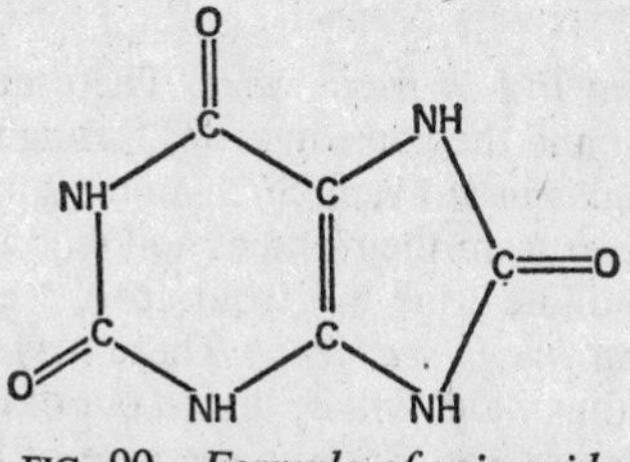

FIG. 90 *Formula of uric acid.*

(*iv*) Other nitrogenous excretory products include the insoluble allantoin and allantoic acid, made use of during embryonic development by amniotes with shelled eggs. There are also endogenously-derived products such as guanine and adenine from nucleic-acid breakdown and creatinine from the creatine of muscles.

21. Mineral ions as waste products. Excess mineral ions taken in with the diet are also excreted by one means or another. In the vertebrates ionic composition of the urine is controlled by endocrines originating in the adrenal cortex.

The important mineral ions of the blood plasma and of the

urine in man and in the seawater crab *Uca* are shown in Table XVI.

TABLE XVI.   MINERAL IONS IN PLASMA AND URINE IN MAN AND UCA

	Ion (concentrations in millimoles/litre)						
	Na^+	K^+	Ca^{++}	Mg^{++}	Cl'	SO_4''	NH_4^+
Man:							
plasma	143	5	5	2.2	103	1	—
urine	143	35	6	9	136	20–60	—
Uca:							
plasma	328	11	16	23	537	42	20
urine	276	16	17	54	622	47	75

It will be seen that though these animals have very different ionic concentrations in their bodies corresponding to the different environments in which they are found both tend to excrete an excess of ions in the urine.

22. Water as an excretory product.

(*a*) *Animals that live in fresh water*. These are *hypertonic* to their environment and thus take in water through permeable surfaces such as the gills or gut wall. Such animals have the problem of eliminating water from their bodies and tend to produce large quantities of very dilute urine, e.g. trout, frog.

(*b*) *Animals that live in sea water*. These may be *hypotonic* to the medium and thus water will be drawn out of them by osmosis across permeable surfaces. Such is the case of marine teleosts, e.g. the herring. Animals in this situation drink quantities of sea water and excrete the contained salt through special cells in the gills. The urine is scanty and concentrated.

Many marine animals are *isotonic* to their environment so have little or no osmotic work to do, e.g. crabs. In the case of marine elasmobranches, e.g. dogfish, retention of urea in the blood increases the osmotic pressure of the organism to that of sea water.

(*c*) *Animals that live on land*. These tend to lose water by evaporation from respiratory and other permeable surfaces. Water conservation is thus of very great importance. Many successful terrestrial groups such as reptiles, birds and insects use uric acid and urates as nitrogenous wastes as these can be excreted in semi-solid form thus conserving water.

NOTE: Some desert mammals excrete urine as much as two and a half times the concentration of sea water but others, such as man, cannot produce a very hypertonic urine and have to lose considerable quantities of water to get rid of nitrogenous waste.

THE EXCRETORY ORGANS OF MAMMALS

23. The relationship of the blood supply to the nephrons. The excretory unit of the mammalian kidney is the nephron (*see* Fig. 91). This term includes the Malphigian body and tubules. There may be several million nephrons in the paired kidneys of a mammal.

Excretory material is derived from the blood which also takes up useful substances such as sugar and certain salts and variable amounts of water. An understanding of these functions is closely related to knowledge of the relationship of the blood vessels to the parts of the nephron. The supply may be considered as follows:

(*a*) The renal artery enters the kidney at the pelvis and divides up into the arcuate arteries. These in turn give rise to many interlobar arteries from which the glomeruli of the nephrons arise.

(*b*) The glomeruli start with a wide afferent vessel which runs into a looped capillary system of ever-decreasing bore. This decrease in vessel size causes a filtration pressure to develop within the glomerulus.

(*c*) The vasa recta consists of a number of looped vessels which derive from the efferent capillaries of the glomerulus. The vasa recta is in close proximity with the loop of Henle and the collecting ducts.

(*d*) Blood from the vasa recta collects up into interlobular veins and thence, via arcuate veins, returns through the renal vein to the vena cava.

24. Function of the nephron. The processes by which the nephron removes waste as urine from the blood without losing at the same time valuable small molecules, water and ions are as follows:

(*a*) *Ultrafiltration.* This takes place across the walls of the glomeruli and substances in the blood of low molecular weight pass across into the capsule. These would include water, salts, glucose and the various nitrogenous wastes of the body. The

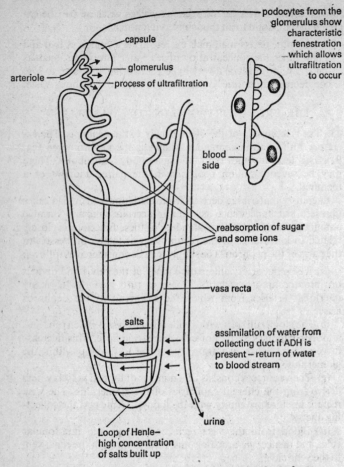

FIG. 91 *The kidney nephron and its blood supply.*

total glomerular surface is very large and it is estimated that some one hundred and eighty litres are filtered across it in twenty four hours. As only a few litres of urine are actually excreted in the same time it is clear that most of the water is circulated in the nephron. This circulation of water through an excretory system is commonly found.

(*b*) *Reabsorption of useful substances.* The walls of the tubules are lined with epithelial cells that have "brush borders" which greatly increase their surface areas. They also contain many mitochondria and these two features together suggest that the tubule cells are actively secreting or assimilating against a concentration gradient.

Glucose is taken up and returned to the blood. The presence of glucose in the urine is a serious sign of malfunction, e.g. diabetes. Mineral ions may or may not be taken up according to hormonal control, exercised mainly by the mineralocorticoids of the adrenal cortex.

(*c*) *Water content of the urine.* According to current theory Na^+ is secreted from the ascending arm into the vasa recta and back into the descending arm. High concentrations of these ions (together with appropriate anions) build up in the region of the loop of Henle both within the tubule and the vasa recta.

The blood is also in close connection with the collecting ducts and if anti-diuretic hormone is present in the blood stream the cells of these ducts become freely permeable to water.

Water is thus drawn out of the passing fluid by osmosis and returned to the circulation. Under such conditions a hypertonic urine is formed. If, on the other hand, the body is hydrated no ADH will be secreted from the pituitary and the collecting duct cells will be impermeable. A weak urine (hypotonic) with large amounts of water will be formed and expelled from the body.

NOTE: The longer the loop of Henle the more the nephron is able to concentrate the urine. Desert mammals have very long loops and can produce urine which may be several times the concentration of sea water. Man with a comparatively short loop is not able to make urine even as concentrated as sea water and thus dies from osmotic and ionic disruption if such is his only source of water.

The method of water regulation described here is called the *counter-current multiplier hypothesis.* There is much experimental evidence that supports it.

(*d*) *Concentration of nitrogenous wastes.* The concentration of urea in human urine is some sixty times greater than that in the blood. Ammonia, present in the blood at only 0·02 mg per 100 ml, is four hundred times more concentrated in the urine, while the creatinine is one hundred times more concentrated.

25. Hormonal regulation of urine composition.

(a) *Anti-diuretic hormone* (ADH), produced in the hypothalamus. It is passed down the hypophysial tract to the posterior pituitary, whence it is secreted into the bloodstream. If the latter shows a tendency to become viscous the hormone is released and its target area is the collecting ducts of the nephrons. As described in **24** it increases the permeability of these to water, thus causing a more concentrated urine to be excreted and more water to be returned to the blood.

(b) *Aldosterone*. This is a mineralocorticoid hormone produced from the adrenal cortex. In its turn its release is determined by pineal and pituitary controls. Aldosterone acts on the tubule, making it more active in its uptake of Na^3, so that a stronger osmotic pull results and more water is returned to the blood. It also has the effect of reducing sodium ion loss.

(c) *Angiotensin*. The precursor of this is found in the bloodstream, but conversion to the active form is triggered off by a chemical from the kidney cortex itself, Angiotensin works on the adrenals stimulating aldosterone release, with the results described above.

(d) *Parathormone*. Made by the parathyroids (very small endocrine glands adjacent to the thyroids) this hormone controls release of phosphate ions from the tubule.

The picture that emerges (and the list above is only a selection of major hormones) is clearly very complex and it can be understood that the composition of the urine at any one time is under multiple homeostatic controls that lead to conservation of the body's fluids and salts at very precise levels.

26. pH control by the kidney.

The kidney helps to maintain the neutrality of the body by secretion of H^+ where the blood has become acid and of OH' where it is alkaline. These ions are carried and secreted in the form of bicarbonate and phosphate buffers, as well as by the amino acid glutamine. Active secretion takes place in the distal convoluted tubule which ensures that the blood returning to the circulation has excess of acid or alkaline ions removed. In renal failure the whole body fluid pH may show fatal fluctuations from normal.

THE EXCRETORY ORGANS OF EARTHWORMS

27. Excretory systems in the earthworm.

There are various excretory systems in the earthworm and, although it lives on land, it is in some ways more like a fresh-water animal as far as excretion is concerned. The systems to be described are the calcareous cells, the chlorogogen cells of the gut and the nephridia.

28. The calcareous cells. These are found in segments 11 and 12 of *Lumbricus* and are pouched diverticula from the oesophageal region of the gut. Carbonic acid, formed from respiration is secreted out of the blood at the same time as Ca^{++}. The former dissociates in the presence of carbonic anhydrase and the Ca^{++} and CO_3'' combine to make insoluble particles of $CaCO_3$. This calcium carbonate is precipitated into the lumen of the gut. Removal of the calcareous gland leads to rapid fall in the pH of the body fluids and there can be little doubt that the function of the calcareous cells is to maintain the acid–alkali balance of the earthworm.

29. The chlorogogen cells. These surround the gut muscles and thus lie inside the coelomic cavity. They are large brown or yellow cells whose role is somewhat analogous to that of the mammalian liver. Within the chlorogogen cells amino-acids assimilated from the gut are deaminated and various nitrogenous excretory substances are formed. These include *ammonia*, *urea* and *uric acid*, and are released into the coelomic cavity. The cells store glycogen and lipids.

30. The nephridia. The most important excretory organs of the worm are its nephridia which are found in pairs in all segments except the first five.

(*a*) *Anatomy of the nephridium.* The nephridium is a tubular structure which opens at the nephridiostome into the coelomic cavity and to the exterior by the nephridiopore.

The mouth of the nephridiostome and the tube into which it leads are ciliated and the beating of these causes a current to flow along the tube. At first this is narrow and much coiled but further down it increases into a middle tube of greater width. This, in turn, leads to the wide tube, which is not coiled. All along its length the nephridial tube is provided with a dense network of capillaries. At the end of the wide tube there is a "bladder" and this leads to the exterior via the sphincter muscle of the nephridiopore.

(*b*) *Ionic regulation by the nephridium.* A mixture of mineral ions, water, nitrogenous substances and even minute particles of excretory matter from disintegrated amoebocytes is wafted into the mouth of the nephridium. From the blood capillaries further water and salts and nitrogenous materials are actively pumped into the narrow tube, and the mixture passes along through the middle tube and enters the wide one. Here any protein and all useful ions and other substances are re-absorbed.

(c) *Water regulation by the nephridium.* Water uptake may or may not take place in the wide tube. In normal conditions some water is taken up but the worm still excretes a very hypotonic urine just like a fresh-water fish. If the worm is living in dry conditions most of the water entering the wide tube will be re-absorbed and a hypertonic urine is formed. This is stored in the bladder and periodically excreted via the nephridiopore.

(d) *Other sources of water loss and gain.* The bodies of worms have very thin cuticles and are relatively permeable to the entry of water and salts. In dry conditions the worm will become dehydrated by evaporation, while in water it swells up and becomes turgid.

Worms continuously secrete mucus and this accounts for a large proportion of their water loss. The mucus also contains some waste nitrogenous substances.

THE EXCRETORY ORGANS OF INSECTS

31. The excretory system of insects. As with worms there are a number of excretory systems present in insects, such as the cuticle itself, the formation of inactive pigments, and the Malpighian tubules and rectum.

(a) *The cuticle.* On moulting, the whole of the exocuticle is shed which accounts for some thirty per cent of the whole. During its formation various nitrogenous substances may accumulate in this region which are lost at the time of moulting. For this reason the cuticle may be called a "kidney of accumulation" analogous to the allantois of a chick.

(b) *Pigments.* A variety of pigments are derived from excretory substances. Reds and browns may come from tryptophan, greens from the breakdown of cytochromes and haemoglobin, and white and yellows from uric acid.

(c) *The Malpighian tubules.* These are tubular glands which are found at the junction of the mesenteron and hind gut. Like the vertebrate kidney they are lined with cells having brush borders and many mitochondria. A scheme indicating the functioning of the Malpighian tubule and rectum is shown in Fig. 92. It can be seen that as with birds and reptiles uric acid is the main

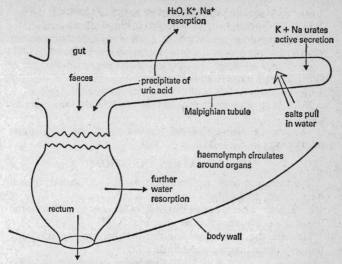

FIG. 92 *The insect Malpighian tubule and its mode of action.*

nitrogenous waste and that a semi-solid urine is formed. With their proportionately very large surface areas and tracheal system, which when fully opened is a major source of water loss, insects have very difficult problems of water conservation. The production of a semi-solid urine and faeces assists the process.

NOTE: Fresh-water insects, such as the larvae of dragonflies, produce copious dilute urine which contains ammonia.

PROGRESS TEST 9

1. What is meant by "homeostasis"? **(1)**
2. Give three examples of homeostatic regulation in man. **(1)**
3. How is the heart beat co-ordinated? **(2)**
4. List the substances and approximate amounts of substances carried by the blood. **(4–7)**
5. Distinguish between lymphocytes and monocytes. **(9)**
6. How may immunity to a disease be acquired? **(11)**
7. Why is it difficult for small animals to regulate their temperature? **(15)**
8. What information is fed into the hypothalamus about body temperature? **(17)**
9. Distinguish between secretion and excretion. **(18)**

10. What are the main nitrogenous wastes of animals? **(20)**

11. List the major elements present in blood plasma. **(21)**

12. If an animal produces large amounts of hypotonic urine in what sort of environment would you expect it to be found? **(22)**

13. Describe the blood supply to the nephron. **(23)**

14. What is ultrafiltration? **(24)**

15. How does ADH control urine concentration? **(24)**

16. Where does water uptake take place in the nephridium? **(30)**

17. Draw a diagram of the functioning of the Malpighian tubules in an insect. **(31)**

EXAMINATION QUESTIONS

The figures in **bold** type indicate the marks allocated to each question or part question.

1. Explain what is meant by the term *homeostasis*. What is the importance of homeostasis for an animal? **(7)**

Describe the various mechanisms that help to regulate body temperature in mammals and birds. **(11)**

Cambridge specimen paper, 1975

2. Describe briefly the action of the mammalian heart as it pumps blood around the body.

An isolated heart, if kept in an appropriate nutrient fluid, will beat regularly for a long time. How is the heart beat maintained in the absence of outside nervous control?

How does the mammalian heart adjust its rate and output per beat in response to changes in body activity?

Cambridge, 1976

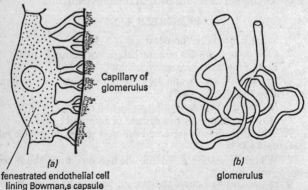

Capillary of glomerulus

(a)
fenestrated endothelial cell lining Bowman,s capsule

(b)
glomerulus

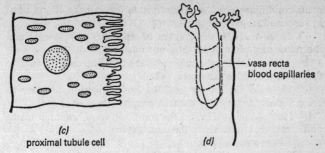

(c)
proximal tubule cell

(d)

loop of Henle and vasa recta capillaries

vasa recta
blood capillaries

3. In a mammalian kidney the following types of cells and structures are to be found.

(*a*) For each cell or structure, anatomy and functioning are very closely connected. Describe the relationship between structure and function in each case.

(*b*) Why do desert animals have very long loops of Henle?

(*c*) List *three* ways in which the structure of the bladder is related to its functions.

Associated Examining Board, 1976

4. The following table gives data about the composition of blood plasma and urine for a man.

	Plasma	Urine
Water	90%	96%
Proteins	7%	0%
Hormones	trace	trace
Ions	0·9%	Variable, usually more sodium, chloride, sulphate than in plasma
Waste substances	trace	2%
Glucose	0·1%	0%
*p*H	7	varies, usually 6

(*a*) What do you deduce about the function of the kidney from information provided by the table? Illustrate your answer by reference to specific differences in the composition of plasma and urine. **(6)**

(*b*) Assuming that renal function is normal, what are the *two* main conditions which are responsible for varying the mineral ion concentration of the urine? **(2)**

(*c*) Given that the salt content of sea water is 3 per cent and the maximum content of salts in human urine is 2·2 per cent, attempt to explain why a shipwrecked sailor is unable to survive for long by drinking sea water. **(2)**

(*d*) How do the kidney and the brain function together to bring about a stable osmotic environment in the body? **(4)**

(*e*) How would you expect the composition of human urine to differ from that indicated in the table above:

(*i*) after strenuous exercise,

(*ii*) during a high protein diet? **(6)**

Associated Examining Board, 1977

5. Write a short account of immunity.

after London, 1975

6. Make large, fully labelled diagrams to show (*a*) the general internal structure of a mammalian kidney, and (*b*) the detailed structure of a nephron.

Explain how the kidney controls the water content and the urea content of the body.

London, 1975

7. Review the mechanisms by which the waste products of metabolism are eliminated from organisms.

Oxford and Cambridge, 1974

8. Discuss homeostasis with respect to regulation of body temperature.

after Oxford and Cambridge, 1975

9. Explain what is meant by active transport. **(3)** Give evidence for the occurrence of active transport. **(5)** Make a labelled diagram of a nephron, indicating regions of intense and active transport at the points you have indicated. **(8)** What features of nephron cells might suggest a transport function? **(4)**

Total **(20)**

Oxford and Cambridge, 1977

10. Give an account of the processes and mechanisms involved in the regulation of the body temperature in mammals.

Oxford and Cambridge, 1976

CHAPTER X

Reproduction and Growth

REPRODUCTION GENERALLY

1. General features of biological reproduction. Reproduction is the process by which living organisms multiply and *it is a characteristic activity of living things*. Reproduction is necessary to make up the losses that occur due to competition, predation, disease, starvation and other causes of death. In most organisms, though not in certain lower ones, reproduction is also necessary to renew the genetic material which appears to have an inborn process of ageing.

A high reproductive rate also allows a species to colonise new environments and, as is clearly seen in plants, the process is connected with dispersal.

Because of the particular form of cell division (i.e. *meiosis*) that occurs in certain forms of reproduction and because of the mixing of genetic material from two individuals, sexual reproduction often results in the production of variations in the progeny. On this variation the process of natural selection can operate.

In most lower plants and some primitive animals such as the Protozoa and coelenterates, reproduction is associated with the formation of resistant spores or cysts which allow survival over unfavourable periods.

2. Sexual reproduction.

(*a*) *The process of sexual reproduction*. This involves the following:

(*i*) The production of haploid sex cells or gametes. In the male these are termed *spermatozoa* and in the male plant *antherozoids*. For the female animal the gametes are called *eggs* or *ova* while in the plant they are *oospheres*.

(*ii*) Two separate sexes (dioecious) are necessary or organisms of both sexes (hermaphrodite).

(*iii*) The fusion of male and female sex cells occurs and produces a diploid organism initially termed the *zygote*.

(b) *The genetic significance of sexual reproduction.* Reduction, division, or meiosis of a diploid organism to form haploid gametes involves the crossing-over of genetic material between homologous chromosomes. This in turn results in new combinations of genes being passed on to the offspring. Advantageous mutant genes may be spread throughout a population by means of sexual reproduction.

(c) *Sexual selection.* This is a part of the process of natural selection. It operates on the large numbers of gametes produced by an individual, especially by the male, to ensure only biologically normal sex cells fuse to make the zygote.

Sexual selection also operates on the whole organism in the sense that mating involves a process of selecting a mate. Intraspecific competition between males to secure a female, courtship and display ceremonies are common preludes to the process of reproduction in higher animals.

In all organisms sexual reproduction normally involves more hazards than asexual, although the long-term effects of the former are necessary to the evolution of the species.

(d) *Parental care.* In mammals, some birds and insects, certain fishes and various other animals, the process of reproduction has become associated with *parental care*. This relationship between parent and offspring may involve some form of communication and in the case of birds and mammals has played an important part in the evolution of their behaviour patterns. Extended parental care and communication and learning were critical factors in the development of the human species.

3. Asexual reproduction. This is a means of reproduction whereby a single individual can give rise to offspring by mitotic divisions of a part of its own body. In a strict sense the process is called asexual if specific bodies are produced such as spores. Where a part of the parent breaks off to produce a new individual the term *vegetative propagation* is used.

Asexual and/or vegetative reproduction is a feature of lower organisms and all plants. It allows rapid colonisation of an environment from an established parent. The process is clearly much less hazardous than sexual reproduction but the offspring produced will have an identical genetic composition.

Higher plants, for example potatoes, strawberries, etc., combine both vegetative and sexual methods of reproduction with considerable success. Specialised stems such as rhizomes, tubers, corms and leaf bases as seen in bulbs are all organs of vegetative propagation.

REPRODUCTION IN PLANTS

4. Reproduction and the life cycle. An outline of the life cycles of representative plants has been given in I.

The general line of somatic evolution of plants has been from unicellular to filamentous multicellular, e.g. *Spirogyra*, and from the latter via branched filamentous and parenchymatous forms, e.g. *Cladophora* to the development of a thallus, e.g. *Fucus, Pellia*.

In higher plants the body is differentiated into root, stem and leaf systems.

Parallel with these changes in the somatic organisation of plants is an evolution of the sexual processes and life histories.

5. Differentiation of gametes.

(*a*) *Isogamy*. This is the simplest type where the two gametes are identical morphologically although there may be different physiological strains, which encourages outbreeding. The conjugation of *Spirogyra* and the sexual reproduction of most species of *Chlamydomonas* are examples of isogamy.

(*b*) *Anisogamy*. The gametes are unequal in size and the larger and smaller fuse to make the zygote. An example is *Chlamydomonas braunii*.

(*c*) *Oogamy*. This is the most advanced form of differentiation of gametes. It is found in the majority of organisms. Oogamy is characterised by a large number of small motile male gametes being produced and a smaller number of large sessile female gametes.

6. Development of alternation of generations.

(*a*) *Simple life histories*. These are seen in some of the algae which have a haploid plant which produces haploid gametes by mitosis. These fuse to give a diploid zygote which undergoes meiosis on germination. *Spirogyra* has such a life cycle (*see* Fig. 93).

There are many advantages in maintaining the diploid state as it allows different alleles to be present (called heterozygosity) and gives the plant a greater potential variability. Unfavourable genes can be buffered in diploid organisms by the dominant allele.

The introduction of genetic material from another individual

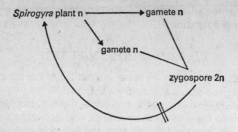

FIG. 93   *The life cycle of* Spirogyra.

This is a haploid plant.

during sexual reproduction also greatly adds to variation among the offspring.

Within the algae themselves as well as in the major advance of the plant kingdom on to the land, there is, associated with the maintenance of the diploid state and the encouragement of outbreeding, the phenomenon of *alternation of generations*.

(*b*) *Alternation of generations*. This phenomenon is most clearly seen in the bryophytes and pteridophytes and the life cycle of these groups may be seen in Fig. 94.

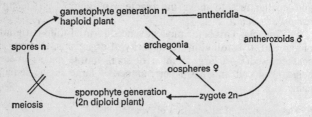

FIG. 94   *The life cycle of bryophytes and pteridophytes.*

These show alternation of generations and the increasing importance of the diploid sporophyte generation.

In the bryophytes the gametophyte generation is dominant but the sporophyte shows a number of pre-adaptations to a successful life on land and in the pteridophytes it is this generation that predominates in the life cycle.

(*c*) *Life cycles of the seed plants.* The seed plants are thought to have evolved from the pteridophytes; several alternative theories have been put forward as to through which route they evolved. Seed plants, or spermatophytes, show the same phenomenon of alternation of generations as seen in the lower plants but the sporophyte (i.e. the plant as we see it) produces two types of spores:

(*i*) *Microspores* (pollen grains) produced in the microsporangia (anthers) give rise to a vestigial male gametophyte and two male nuclei. One of these fertilises the female nucleus and the other joins with the secondary nucleus to make the triploid endosperm nucleus (this is true only of angiosperms).

(*ii*) *A megaspore* produced within each megasporangium and the megaspore produces a vestigial female gametophyte (the embryo sac) and a single female sex cell or nucleus.

Figure 95 shows the life cycle of the seed plant, in terms of its

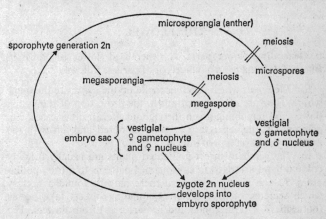

FIG. 95 *The life cycle of the seed plant.*

Working out the homology of this life cycle with those of lower plants was a considerable achievement of comparative botany and is strong proof of evolutionary descent.

homologies with the cycles of lower plants.

7. Advantages of the seed habit. The seed habit has many advantages over the cycles of lower plants and the spermatophytes are the dominant plants on earth. Advantages of the seed habit are these:

(a) The development of pollen grains which may be transferred by wind or animals (usually insects) from one plant to another. These replace the free-swimming antherozoids which restrict the successful reproduction of non-seed plants.

(b) The female gametophyte is retained on the sporophyte where it is protected and nourished.

(c) After fertilisation the whole female gametophyte and the surrounding integument becomes the seed. The seed is adapted for the dispersal and survival of the plant in unfavourable conditions. It carries a much larger supply of nutrient material than the spores of lower plants.

(d) In angiosperms (as distinct from the gymnosperms such as conifers) the seeds are surrounded by a modified ovary wall, the *carpel*, which often plays an important part in successful dispersal of the former.

THE FLOWER

8. Generalised flower parts. A generalised flower is shown in Fig. 152. It is made up of whorls or of parts as follows:

(a) *The sepals* which together make up the *calyx* are the outer whorl. They are often important in the protection of the young flower bud remaining green throughout life. In other cases, e.g. delphinium, the sepals are brightly coloured and act with the petals in the attraction of insects.

(b) *The petals* of insect-pollinated plants are usually brightly coloured and arranged in conspicuous patterns to attract pollinators. At the base of the petal there may be found a nectary which secretes sucrose nectar. Pollinating insects take nectar (butterflies) or both pollen and nectar (bees) from flowers. The petals make up the *corolla*.

(c) *The stamens* together make up the *androecium* or male part of the flower. The anthers are borne at the end of the stamen on long filaments. The development of anthers and micro-sporanga may be seen in Fig. 96.

(d) *The carpels* which constitute the *gynoecium* or female part of the flower consist of an ovary at the base with a variously developed style and terminal stigma. Carpels are homologous with megasporophylls and the ovules they contain are megasporangia. Within the ovule a single megaspore develops and this, in turn, produces the vestigial female gametophyte (the embryo sac) with

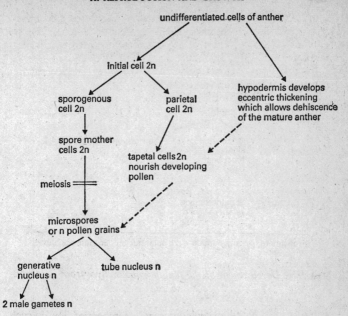

FIG. 96   *Scheme showing the development of the anther and micro-sporangia and microspores (pollen grains).*

its single oosphere. The development of the carpel may be seen in Fig. 97.

8. Pollination mechanisms and agents. The transfer of pollen from the anther to the stigma of the same flower or to another flower of the same species is called pollination. It is normally brought about by wind or insects and most flowers show a variety of adaptations to one or other of these pollinating agents (*see* Table XVII).

9. Means used for reducing or preventing self-pollination. One of the biological objects of sexual reproduction is to produce *genetic variation* by combining genes from different sources. Most flowers are hermaphrodite (exceptions are hazel, campion, cucumber, maize, etc.) and many elaborate mechanisms are used to reduce or prevent self-pollination. A few of these are described below:

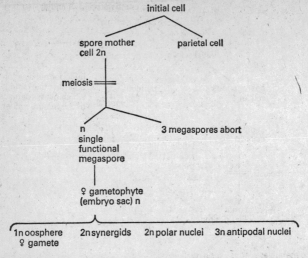

FIG. 97 *Development of the female gametophyte from the mega-spore.*

(a) *Differential maturation of parts.* Numbers of flowers such as deadnettles and buttercups have the stamens maturing before the carpels. This decreases, at least for a period, the chance of self-pollination.

$$\text{protandny} = \text{maturation of } \male \text{ before } \female$$
$$\text{protogyny} = \qquad \text{,,} \qquad \text{,,} \female \quad \text{,,} \male$$

(b) *Physical separation of parts.* This may take many forms. In primroses some flowers called "pin-eyed" have long styles and short stamens half way down the corolla tube, others, the "thrum-eyed", have short styles and stamens at the top of the corolla tube. In the daisy family the receptive surface of the stigma is kept closed while the style grows through the ring of stamens. Leguminous plants such as gorse have a complex trigger mechanism whereby the pollinating insect causes the discharge from the keel of the stigma followed by a mass of pollen. Self-pollination is prevented by a circlet of hairs.

(c) *Self-incompatibility.* Some plants, especially grasses, may have sets of mutually incompatible genes, such that the pollen will not germinate or else grows very slowly on the carpel.

TABLE XVII. COMPARISON OF INSECT- AND WIND-POLLINATED FLOWERS

	Features connected with insect pollination	*Features connected with wind pollination*
Common features	Homologous floral parts. A pedicel or stalk which carries the flower above the leafy part of the shoot.	
Calyx and corolla	Conspicuous coloured petals and/or sepals. Colours that contrast with green are in the range seen by insects. Red and white flowers may reflect U.V. light which is visible to bees.	Small green glumes, lodicule, etc., homologous with the perianth.
Scent and nectar	Production of scent and nectar and markings act as honey guides.	No scent or nectar.
Androecium	Anthers produce moderate amounts of large sticky pollen grains. Filaments not longer than corolla.	Anthers produce very large amounts of small dry pollen. Filaments very long and anthers pendulous which makes them shake in the wind and discharge pollen.
Gynoecium	Style and stigma contained within corolla.	Large feathery style for catching pollen grains.
Self-pollination	Various means for reducing self-pollination.	Genetic incompatibility common between pollen and carpel on the same flower prevents self-fertilisation.
Examples	Bean, buttercup, deadnettle.	Grass flower.

SEED AND FRUIT FORMATION

10. The events following pollination in angiosperms. After the pollen has been transferred either to the stigma of the same flower or to that of another flower of the same species it germinates and develops a long pollen tube. This tube grows down the style and crosses the cavity of the carpel, finally entering the micropyle of the ovule. Within the germinated pollen grain there are a generative and a tube nucleus. The former gives rise to two male nuclei (*see* Fig. 98).

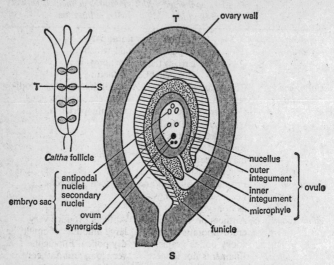

FIG. 98. *Transverse section of a mature carpel (from a delphinium follicle).*

After the tube has entered the micropyle the two male nuclei pass into the embryo sac, one fusing with the ovum to form the zygote and the other forming a triploid nucleus with the two secondary nuclei.

11. Development of the embryo and the endosperm. The zygotic nucleus rapidly divides and forms a chain of cells called the suspensor which has the effect of pushing the developing embryo well into the centre of the embryo sac. The embryo then begins to differentiate a shoot or plumule, a root or radicle and one or two cotyledons according to the species.

Meanwhile the triploid endosperm nucleus has been rapidly dividing and cutting off large cells which obtain food from the nucellus surrounding the embryo sac.

There comes a point when the endosperm tissue fully occupies the embryo sac with the exception of the small embryo itself.

What happens next again depends on the type of plant; in endospermic plants such as maize and cereals the food store remains in the endosperm while in non-endospermic plants (the majority of dicotyledons) the food is transferred to the cotyledons which become very large, e.g. bean.

12. The seed and the fruit. Hormones produced during embryo development cause other changes to occur in the carpel.

The integuments surrounding the ovule harden and become the impermeable testa so that the ovule has now become the seed. The carpel wall changes in a variety of ways and forms the fruit which in most cases assists the dispersal of the seeds. The simplest form of fruit is where the carpel wall hardens and the fruit is described as an achene. In other cases the carpel wall, or pericarp, changes into an outer coloured skin, flesh and a hard stone around the seed. Such fruits are succulent and animal dispersed. The pericarp may also become expanded to catch the wind, e.g. the sycamore.

13. The germination of seeds. While it is well known that seed germination depends on such factors as a period of dormancy, suitable temperature, oxygen, water and sometimes light, other factors are now seen to be involved.

Many seeds besides having an impermeable coat also contain *inhibitor* substances such as *abscissic acid*. This latter prevents action of *auxins*, *gibberellins* and possibly *cytokinins* as well, all of which are plant hormones involved in growth and differentiation (*see* XI, **40–46**).

The inhibitor substance may need to be leached out of the testa before germination can begin. When its level has been reduced to a critical point it appears that the embryo (as in barley for example) secretes gibberellic acid. This interacts with the embryo itself and stimulates protein synthesis. It also diffuses out to the endosperm, in particular the aleurone layer below the testa, and causes release of hydrolytic enzymes.

The mitochondrial tricarboxylic acid cycles, previously inactive are switched on and the respiration rate of the seed rises several thousand times.

Thus a complex interplay of both external factors and internal ones controls the onset of germination and ensures that this will only occur when environmental conditions are favourable for survival.

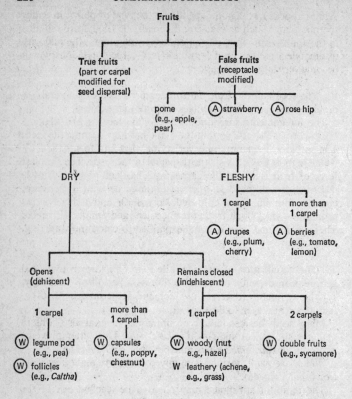

FIG. 99 *Classification of fruits in angiosperms.*

Types of fruit are useful for the identification of species.

14. A classification of fruits. The many modifications of the carpel for dispersal of the seeds is shown in Fig. 99. In some cases such as strawberries, apples and rose hips the succulent part of the "fruit" is in fact an enlarged receptacle.

REPRODUCTION IN MAMMALS

15. Reproductive cycles in mammals. Many mammals have a breeding season adapted to the climatic conditions such that the young are reared in the most favourable part of the year. During this season the female may have a number of oestrus cycles (corresponding to the menstrual cycle of the human female) when ovulation occurs. Certain mammals such as man and many rodents have regular oestrus cycles throughout the year and can therefore reproduce at any time.

Once sexual maturation has taken place under the influence of the pituitary, thyroid and sex hormones, the onset of oestrus is partly determined by environmental change and partly by an innate rhythm within the animal. Increasing daylight in the spring is an important stimulus to mammals that breed in temperate climates.

Sex cycles are also present in the male mammal and lead to the behaviour becoming orientated towards reproduction at certain times of year. The female in oestrus or "on heat" acts as the immediate stimulus for mating to occur and the female is also receptive of the male at this time.

Out of the breeding season the majority of mammals do not show any sexual behaviour and the gonads themselves may decline as does the level of the circulating sex hormones.

16. Sex organs and activity in the male.

(*a*) *Hormones.* Follicle-stimulating hormone from the pituitary causes development of the semeniferous tubules of the testis, while the interstitial cell-stimulating hormone (ICSH, the equivalent of lutenising hormone in the female) causes testosterone to be produced by the testis. Testosterone causes the secondary sexual changes such as enlargement of the sex organs, increase of muscle power, metabolic changes and sex drive.

(*b*) *Sperm development.* The process of male gamete formation or spermatogenesis has been described (*see* V, **39**). This process occurs within the semeniferous tubules of the testis (*see* Fig. 100) and millions of mature spermatozoa accumulate in the lumen of the tubules. The mature sperm in man is some 2μm long. It has a large nucleus full of gene material and from the base of this emerges the axial filament of the tail. The region directly behind the nucleus is thickened into a sheath full of mitochondria which provide ATP for the lashing movements of the tail.

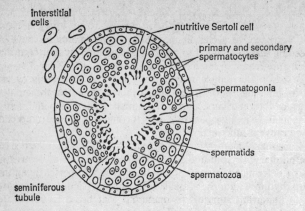

FIG. 100   *A semeniferous tubule from the mammalian testis.*

(c) *Accessory glands.* The sperm may be stored in the coiled tubes of the vasa efferentia or there may be a seminal vesicle at the top of the vas deferens. Accessory glands such as the prostate and Cowper's glands secrete various substances that make up a part of the semen. These include alkali against acid urinary excretion, fructose sugar, which is taken up directly by the sperm and hormones, which tend to depress sperm activity while still in the male.

(d) *The penis and copulation.* For the process of internal fertilisation which is a part of the mammals' effective colonisation of the land an erectile organ called the penis is present. On sexual stimulation the spongy corpus cavernosa spongiosum of this organ becomes full of blood so that it becomes rigid and can be inserted into the vagina of the female. In man a single ejaculation contains some five hundred million sperms.

17. Sex organs and cycles in the female. The sexual co-ordination of the female is more complicated than that of the male as she plays a more important role in the process of reproduction.

(a) *The ovary and the oestrus cycle* (for the human female). There are two ovaries which liberate eggs into the Fallopian tubes and thence via the oviducts to the uterus. Surrounding the ovary is the germinal epithelium from which large numbers of tiny follicles are budded off during embryonic development. Particular follicles are stimulated to mature by follicle-stimulating hormone (FSH) from the pituitary. A central oogonium forms in the follicle and this enlarges and gives rise to a primary oocyte. The first stage of meiosis produces a secondary oocyte and the second stage the ovum (*see* V, **39** for details of oogenesis). The ovum is some 0·2 mm diameter in humans.

During its development the follicle produces oestrogen which feeds back and causes luteinising hormone to be secreted by the

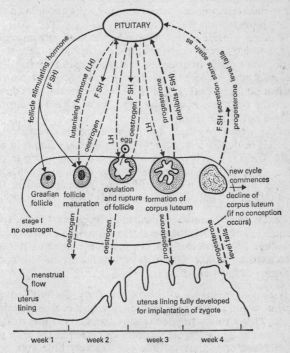

FIG. 101 *Hormonal activity co-ordinating the sex cycle of the human female.*

pituitary. At the end of the second week of the cycle ovulation occurs and the mature egg is discharged (*see* Fig. 101).

The lining cells of the ruptured follicle are stimulated to develop by the LH and form the corpus luteum. By the third week they are actively secreting the hormone progesterone. This hormone prevents further liberation of FSH and causes other changes in the female body associated with pregnancy. The walls of the uterus are now at their maximum size and ready for implantation of the zygote, should fertilisation occur.

If fertilisation does not occur the corpus luteum declines as does the supply of progesterone. FSH activity then starts off a new cycle. The menstrual flow is the shedding of the uterus lining.

(*b*) *The uterus*. This is a double structure in primitive mammals such as rodents but fusion occurs in many types and in man the uterus is a single structure.

The uterus is made of smooth muscle and lined with an endometrium which becomes enlarged for the reception of the zygote. The organ is capable of extension to something like five hundred times its normal volume. A hormone called oxytocin, from the posterior pituitary, causes rhythmical contractions to pass down the uterus. The uterus opens into the vagina which is a distensible tube leading to the exterior.

(*c*) *The mammary glands*. Oestrogen hormones circulating in the blood cause the enlargement of the mammary glands in the thorax. There is a cycle of activity of these corresponding to the oestrus cycle and during pregnancy they develop complicated secretory ducts. The pituitary hormone prolactin causes the actual secretion of milk shortly after the birth of the offspring.

18. Fertilisation. Sperm are deposited in the vagina during copulation and swim up into the uterus. Fertilisation normally occurs somewhere in the oviduct and the first sperm entering the egg causes a change to take place in its vitelline membrane which prevents entry of further sperm.

19. Implantation. Shortly after the sperm entry, the fertilised egg or zygote begins cleavage and in a few hours a morula of cells has formed. This develops trophoblastic villi around its outer surface and in this stage the zygote becomes implanted in the enlarged endometrium of the uterus. The egg has very little yolk and nourishment is taken in through the villi from the uterus lining.

THE MAMMALIAN FOETUS

20. The development of the foetus. The mammalian egg is very small and the developing foetus obtains its nourishment from the mother via the placenta. The very primitive monotremes are egg-laying mammals and the marsupials only retain the young inside their bodies for a short period after which it is transferred to an abdominal pouch into which the mammary glands open.

21. The extra-embryonic membranes. While the stages of gastrulation and neurulation and organology are taking place the foetus is also developing a system of extra-embryonic membranes which are important in its protection and nutrition.

From the head and tail regions grow out the amnion on the inside, i.e. next to the foetus and the chorion on the outside. These ectodermal layers are lined with mesoderm and within this latter extra-embryonic coeloms form. The amnion encloses the embryo in a cavity of fluid which is isotonic with its own tissues and gives it support and protection from desiccation and mechanical damage.

From the gut of the embryo projects the yolk sac (which is very small in mammals) and later the allantois forms. This is an endodermal layer lined with splanchnic mesoderm and it grows out to meet the chorion. The region where the chorion and allantois meet becomes the chorio-allantoic placenta and fuses to form a part of the uterus wall.

Both maternal and foetal tissues involved in the placenta are richly supplied with blood and there is a very large surface area where exchanges between the two blood systems can occur (*see* Fig. 102). Substances of small molecular weight such as sugar, fatty and amino-acids, vitamins, ions, water and oxygen as well as some antibodies diffuse from the mother's blood into that of the foetus. Nitrogenous wastes and carbon dioxide pass in the reverse direction.

In some mammals, such as man, the layers involved in the formation of the placenta, especially the maternal ones, have been reduced during evolution allowing for more efficient exchange. Man has a haemo-chorial placenta where all the maternal tissues are lacking. This allows a most effective exchange but also involves considerable blood loss at parturition.

22. Parturition (birth). In the case of man the foetus is practically complete after three months and the remaining six months' gestation is largely a time for growth. Towards the end of this time hormones such as progesterone which have maintained the uterus in its gravid state begin to decline while oxytocin builds up. The latter causes contractions to start in the uterus and as these become stronger the amnion breaks and the whole embryo is forced out of the uterus. The umbilical cord which connects the

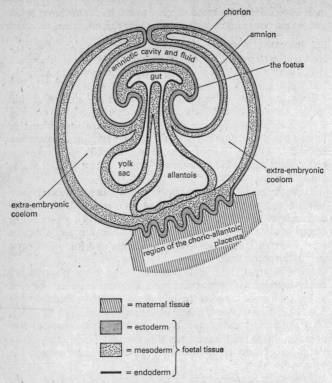

FIG. 102 *The extra-embryonic tissues and the formation of the placenta in the mammal.*

In the next stage the allantois spreads until it occupies most of the extra-embryonic coelom.

baby with the placenta is severed and the baby starts to circulate blood through its lungs and to breathe air.

The placenta is expelled from the body shortly after birth.

23. Changes in the heart and circulation at birth. Blood returning to the foetus from the placental vein enters the liver and passes to the right side of the heart via the vena cava. The foetal heart has a hole, the *foramen ovale* between the right and left auricles and most of the blood passes from the right auricle to the left and thence to the right ventricle. From here it enters the aorta. The small amount of blood leaving the right ventricle starts along the pulmonary artery but at this stage the lungs are collapsed and the blood is side-tracked into the aorta via the open ductus arteriosus.

At birth the umbilical cord is severed and low oxygen pressures in the circulation that immediately result stimulate (via the respiratory centre in the medulla) the taking of the first breath. This

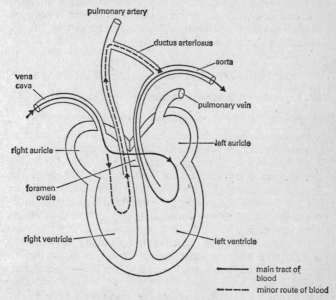

FIG. 103 *Foetal heart and arches.*

All the blood passing through the heart is oxygenated.

in turn leads to an opening of the pulmonary circulation. Shortly after birth the foramen ovale and the ductus arteriosus seal up and a true double circulation is established.

24. Parental care. This is a characteristic of the reproduction of all mammals and is particularly intense and prolonged in man. The offspring are fed from the mother's milk and protected by the aggressive behaviour of the female to intruders.

In some animals such as lions the male contributes to the rearing of the young but in herd animals such as reindeer or sheep one dominant male will sire many offspring but not be paired up with any particular female.

During the period of parental care there is a process of learning by the young and of copying the activities of the parent.

A pair staying together with their offspring for a long period is the basis on which social behaviour arose. Parental care is an extremely efficient form of behaviour that does much to eliminate the hazards to which the young animals are exposed. It is present in both mammals and birds and is a contributory factor to the success of these two groups of animals.

REPRODUCTION IN SOME OTHER ANIMALS

25. Reproduction in selected vertebrates. *See* Table XVIII. p. 237.

26. Special features of reproduction in selected invertebrates.

(*a*) *Amoeba.* Binary fission takes place by mitosis. There is no sexual reproduction. Multiple sporulation may occur.

(*b*) *Paramecium.* These may reproduce asexually by fission. Sexual reproduction involves the micronucleus which forms four haploid gametic nuclei. Interchange of such nuclei may take place with another individual at conjugation. A process of endomixis which is also sexual but involves only a single individual is known to occur.

(*c*) *Hydra.* Asexual reproduction is by budding. Sexual reproduction takes place in unfavourable conditions. Hydra is protandrous and the testes develop well up the body. Sperm from one individual is chemotactically attracted to the ovary of another which contains a single yolk-filled egg. Encystment of the zygote occurs. Sexual reproduction is associated with survival, overwintering and dispersal.

(*d*) *Planaria.* Asexual reproduction by fission may occur. Flatworms are hermaphrodite but cross copulation takes place so the sperms from one worm are exchanged with another and self-fertilisation does not occur. Yolky eggs with shells are laid and the worms swim against the current at the time of reproduction. In connection with its fresh-water environment the free-living larval stage is suppressed.

(*e*) *Earthworm.* There is no asexual reproduction in earthworms although this may occur in the related polychaetes. Earth-

TABLE XVIII.   REPRODUCTION IN SELECTED VERTEBRATES

Animal	Special features of male	Special features of female	Offspring
(a) Herring	Testis uses part of nephric duct to discharge sperm via cloaca. External fertilisation.	Spawn against the current. Some 30,000 eggs laid on sea floor.	Larvae provided with yolk. Join the zooplankton during growth. No parental care.
(b) Dogfish	Internal anatomy as above but pelvic fin modified as claspers for internal fertilisation.	A few very yolky eggs are produced. These are provided with a chitinous egg purse. The latter are secured to sea floor. Urea may be secreted into the egg purse.	No direct parental care but the provision of a quantity of food and protective purse increase chances of larval survival.
(c) Frog	Nuptial pad on thumb Use of cloaca and kidney duct as in the fish. In some species elaborate courtship and secondary sexual characters. Mating in water. External fertilisation.	A large number between 1 and 2,000 eggs supplied with yolk and albumen. Laid in batches. Albumen may put up temperature. There is no parental care.	Tadpole has unpleasant tasting skin. Well adapted to vegetable, then animal diet. Metamorphosis at twelve weeks.
(d) Thrush	Special duct from testis to cloaca. Elaborate nesting and courtship. External fertilisation.	4–5 large yolked shelled eggs laid in specially constructed nest. Incubation and feeding of nestlings. Parental care well developed.	Young born helpless but fed and protected by parents.

worms are hermaphrodite but a reciprocal exchange of sperms takes place and the foreign sperms are stored. A cocoon with a protective layer and filled with albumen is secreted and passed anteriorly over the body of the worm. Into this cocoon the eggs and sperm are deposited. A single young worm emerges from the cocoon. There is supression of the free-living larval stage seen in polychaetes.

(f) *Snail*. There is no form of asexual reproduction in the snail. The animal is hermaphrodite and an elaborate precopulation ritual occurs with the discharge of darts. An ovotestis makes both eggs and sperm but at different times. Foreign sperms are stored in a spermatheca for several weeks before swimming up a common duct to fertilise the eggs. These have albumen supplied and a tough shell and are laid in indentations in the soil. The free-living larval stage characteristic of marine molluscs is suppressed.

PATTERNS OF GROWTH IN LIVING ORGANISMS

27. The growth curve. The growth of both individuals and whole populations tends to follow a *sigmoid curve*. This is one that starts slowly then increases to a maximum rate and then reaches an asymptote where no further change occurs. Mathematically this type of curve can be interpreted by assuming that the rate of growth at any time is proportional to the difference between the present weight and the final weight that will be attained.

In the case of the growth of an individual the first phase is marked by an increase in cell numbers but not necessarily in whole size. This is seen in the formation of a gastrula of hundreds of cells from a single-celled zygote. The phase that follows is known as *the grand period of growth* and here rapid increase in weight of the whole organism occurs. As maturity is reached the growth rate tails off and equilibrium is reached. The final stage of senescence is marked by decrease in weight as breakdown exceeds growth and this terminates in death.

In the case of many plants (especially annuals such as maize) there is a sharp decrease in dry weight following the production of seeds and flowers. This is also true of fishes such as the plaice and salmon.

28. Growth and form. Growth of many marine organisms such as lobsters and seaweeds tends to be unlimited and the organism continues to grow until its death. In land-living species the more demanding nature of the environment produces a limited form of growth and increase of size ceases at a certain stage. It is interesting to note that tree growth is theoretically unlimited (trees of at least 2,000 years of age are known).

(*a*) *Animal forms.* Animals which normally have to move in order to feed have quite definite specific forms and are not dependent on the environment for their shapes, although their sizes may be determined by nutrition and other factors.

(*b*) *Plant forms.* Plants have certain definite regions which can grow, which are the primary and secondary meristems. In this they contrast with animals of which all the organ systems posses their own capacity for growth.

(c) *Regeneration*. Regeneration and wound healing are forms of growth whereby the specific form of the organism is regained after damage. Aquatic animals tend to have very considerable powers of regeneration of parts and have large numbers of undifferentiated cells, e.g. the interstitial cells of hydra. In some cases, e.g. crabs, whole limbs may be lost and regenerated.

29. Factors affecting growth. Whatever genotype the individual may have inherited, its final size and its growth are under the influence of certain environmental factors. They are as below:

(a) *Nutrition*. A principal factor is nutrition for without a sufficient diet normal size cannot be reached. This is clearly seen in the growth of crop plants at different fertiliser levels and the results of malnutrition in animals. Young rats fed on a diet deficient in vitamin A show stunted growth as do children who have insufficient vitamins A and D.

(b) *Light*. The growth of plants is affected by light and an insufficiency causes etiolation. It is not so important in animals although in many mammals sunlight is essential for vitamin D synthesis.

(c) *Temperature*. Another critical factor in growth is temperature. For warm-blooded animals low temperatures tend to increase the life span whereas high temperatures shorten it. Forcing of plants in high temperatures brings about increase in the growth rate.

(d) *Endocrines*. Endocrines in animals and auxins in plants also play an important part in the internal environment that influences the growth rate. Addition of an auxin to a dormant bud can cause new growth to occur and pituitary and juvenile hormone excess and deficiency influences the normal growth pattern of vertebrates and insects respectively.

30. Measuring growth. In order to make quantitative observations on growth it is necessary to measure the process. In terms of population this is easy but in terms of the individual it can be very difficult.

Measurements of increase of height or width or volume can be used to measure growth rate, e.g. the *auxometer* which measures the increase in length of a shoot and furnishes useful information about the response of the plant to variations in light. Nevertheless a plant could show etiolation in the dark and appear on the basis of extension to have grown a great deal.

Generally the increase in dry weight is taken as being the most satisfactory index of growth although this must involve killing a sub-sample of the specimen under investigation in order to obtain the dry weight. If a large number of experimental animals or plants are used this method probably gives the most satisfactory results.

PROGRESS TEST 10

1. How is sexual reproduction defined? **(2)**
2. What are the advantages of vegetative reproduction? **(3)**
3. What is meant by "alternation of generations"? **(6)**
4. Where would the male and female gametophytes be found in seed plants? **(6)**
5. What is a seed? **(6)**
6. What are the features of wind-pollinated plants? **(8)**
7. What is a fruit? **(12)**
8. What is the survival significance of a breeding season in many mammals? **(15)**
9. How is the menstrual cycle co-ordinated in the human female? **(17)**
10. What are the extra-embryonic membranes? **(21)**
11. What is meant by "parental care"? **(24)**
12. What is the "grand period of growth"? **(27)**
13. Which factors may influence the rate of growth of an organism? **(29)**

EXAMINATION QUESTIONS

The figures in **bold** type indicate the marks allocated to each question or part question.

1. Explain why it is **incorrect** to state:

(a) that the pollen grain of the flowering plant is equivalent to the spermatozoon of a mammal, **(6)**

(b) that the anther is the male organ of the flowering plant, and **(6)**

(c) that the ovum of a mammal is equivalent to the ovule of a flowering plant. **(6)**

Cambridge specimen paper, 1975

2. The following account contains many errors of botanical fact and interpretation. Read the passage and identify these

errors. List them, and then re-write the account at a standard appropriate to G.C.E. "A" level work.

"The flower is the sex organ of the plant and unlike most animals, flowers contain sex organs of both the male and the female. The stamens are male and the ovaries female. A flower tries to attract insects by its bright colours and sweet nectar because insects such as bees carry pollen from one flower to another. The pollen brought by the bee grows on the ovary until it reaches the egg which it fertilizes. The egg becomes the seed. Flowers try not to get their own pollen onto the ovary as seeds made from their own pollen are weaker than those formed by pollen from another flower of the same species."

Associated Examining Board, 1973

3. By reference to a bryophyte and to a fern explain what is meant by the term "alternation of generations". Discuss whether similar alternation of generations occurs in flowering plants. **(12, 8)**

London, 1976

4. Discuss the structure of the flower in relation to its functions.

London, 1976

5. Write concise notes on *two* of the following: (*a*) alternation of generations in plants; (*b*) seed germination.

after London, 1976

6. Discuss the advantages and disadvantages of the different systems of reproduction found in organisms.

Oxford and Cambridge, 1975

7. What are the differences between a pteridophyte and a flowering plant over equivalent periods of the life history?

after Oxford and Cambridge, 1977

8. The onset of reproduction is controlled by internal and external factors. Discuss.

after Oxford and Cambridge, 1977

CHAPTER XI

Co-ordination

SYSTEMS OF CO-ORDINATION IN ANIMALS

1. The nervous and endocrine systems. The physiological and behaviour activities of animals are co-ordinated either by the nervous system or by endocrines or, as in such cases as digestion and response to stress, by both nervous and endocrine systems operating together.

As far as is possible to generalise, the nervous system co-ordinates *specific muscular* (*or glandular*) *activity* and the nerves are directly connected to the muscles (or glands) that they control. Nervous co-ordination takes place rapidly. In contrast the endocrine or hormonal system may have very *wide target areas* and the specific chemicals of the system are carried to these areas by the bloodstream. Most endocrines operate over long periods of time but some, for example adrenalin, bring about their effects very rapidly.

Stimuli are transmitted from one nerve to another by means of chemicals such as acetylcholine and in the nervous systems of mammals and other animals are nerve cells which liberate endocrines. These latter are called neuro-secretory cells. It is clear that there is a close homology between the two types of co-ordination systems.

2. Simple and complex nervous systems. It has been suggested that the most primitive type of nervous co-ordination consisted of a stimulus acting on a receptor which was in turn connected directly to an effector which brought about a response. This is a closed system and the stimulus can bring about only the one response, as shown in Fig. 104.

NOTE: A simple arrangement of this sort is seen in the discharge of the nematocysts of coelenterates, although even here other factors such as the nutritional state of the whole animal affects the situation.

The next stage in complexity in evolution was probably the development of an intermediate system of nerves between re-

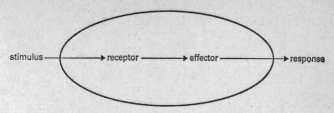

FIG. 104 *Closed nervous system.*

ceptor and effector. Such a central system would allow more flexibility in response as alternate pathways would activate different effectors, as shown in Fig. 105.

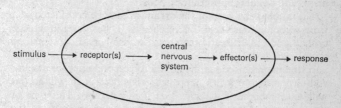

FIG. 105 *Nervous system with intermediate set of nerves between receptor and effector.*

This level of organisation may be found in animals such as the platyhelminthes (flatworms).

Originally, it may be supposed that the central nerve mass was only active when stimulated by the receptor input but at a later stage it began to produce its own inherent activity. Thus rhythmical discharges can be detected from the cerebral ganglia of annelids and all animals above this level of organisation. The functioning and behaviour of animals at this level of evolution thus was partly due to the internal rhythms and activity of their central nervous system which also integrated the peripheral input from the receptors.

This active central nervous system became more elaborate and very complex behaviour patterns and effector actions are found correlated with its increasing size, as shown in Fig. 106.

3. Co-ordination and homeostasis. The concept of homeostasis has been described at the beginning of IX. The maintenance of

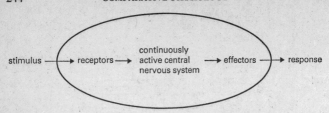

FIG. 106　*Nervous system with active central nervous system.*

the steady state, or homeostasis, of an animal is under the control of its nervous system and endocrines which between them are able to detect changes in the internal and external environments and initiate appropriate responses.

The whole behaviour of the organism may be considered as a form of homeostasis and this reaches its highest level in insects, cephalopod molluscs and mammals, whose nervous systems are most developed.

In the mammals alone the ability to respond appropriately to a situation before it has happened exists (as a man will go out with his umbrella if it seems that it might rain). This type of behaviour is called *intelligence* and it may be regarded as the highest and most sophisticated form of animal activity.

Later in this chapter the differences between intelligence and instinct and other forms of behaviour and the relation these have to the central nervous system will be considered. As an introduction to the way in which the nervous and endocrine systems are put together and operate we will first consider the co-ordination of the mammal.

THE MAMMALIAN NERVOUS SYSTEM

4. The neurone. Neurones consist of a cell body and associated processes that grow out from it.

(*a*) *The cell body.* The cytoplasm of the cell body has minute neurofibrils running through it and out into the processes. It also has many Nissel granules which are made of ribonuclear proteins and are thought to be concerned in synthesis. The nucleus is large and has a conspicuous nucleolus, but no centriole which may be why the mature nerve cell cannot divide.

(*b*) *Nerve processes* (*see* Fig. 107). The two processes leaving

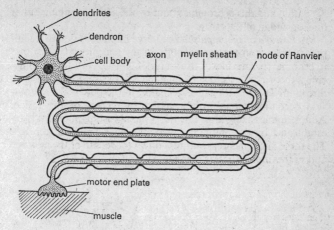

FIG. 107   *A schematised neurone.*

The axon may be over a metre in length, while the cell body will be
situated within the central nervous system.

the cell body are the *axon* and the *dendron* and this latter gives
rise to smaller branches called *dendrites*. Sensory nerves have
long dendrons and short axons while in motor nerves the con-
verse is true.

The cell bodies of neurones are always found in the central
nervous system or in ganglia while the collections of processes or
fibres run together to make up the nerves themselves.

Associated with neurones are the *Schwann cells* and these are
large and envelop the axons and dendrons within their cytoplasm. In medullated neurones the Schwann cell secretes a myelin
sheath of lipid and protein which is important in rapid-conduction
mechanisms.

A cross section of a whole nerve shows that the medullated
and non-medullated fibres enclosed in the Schwann cells are embedded in a connective endoneurium. This in turn is surrounded
by a perineurium and a strong epineurium.

5. The nerve impulse.

(*a*) *The resting potential*. The cell surface membrane or neurolemma is positively charged with respect to the inside by some
70–90mV. This is due to the accumulation of Na^+ outside and of

K⁺ within and this is brought about by the active pumping out of one ion and in of the other.

(*b*) *Passage of the impulse*. The nerve impulse is due to a temporary change in the permeability of the neurolemma such that it becomes permeable to Na⁺ and K⁺. The Na⁺ thus rapidly enters the nerve by diffusion while the K⁺ diffuses in the opposite direction. This in turn causes a change in the potentials, the outside becoming temporarily negative in respect of the inside. Almost immediately after this the original resting potential is re-established by the pumping out of the Na⁺ and in of the K⁺. The same phenomenon continues down the length of the neurone and so the impulse passes along (*see* Fig. 108).

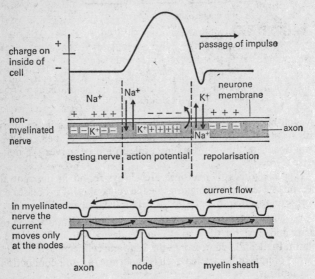

FIG. 108 *The passage of the action potential along the neurone.* The presence of the myelin sheath and its nodes greatly increases the speed of conduction.

(*c*) *Speed of conduction*. This is normally due to the area of cross-section. This applies to invertebrates as well, as giant fibres for rapid conduction velocities are a common feature in arthropods. The giant fibres of the squid have been widely used in research on the mechanism of nervous conduction.

In mammals, A fibres are up to 20μm thick and include medullated nerves, B fibres are from $1-5\mu$m thick and include medullated nerves of the autonomic system while the thinnest and slowest nerves (conduction velocities 1 metre per second) are non-medullated and include certain sensory and autonomic nerves.

As described above, medullated nerves have a myelin sheath secreted by the surrounding Schwann cells and this is periodically interrupted by the *nodes of Ranvier* where the surface of the axon is exposed. In such nerves conduction is saltatorial and velocities can be reached without the large cross-sectional area necessary for fast non-medullated nerves. Some of the peripheral medullated nerves in mammals will conduct at up to 120 m per second.

(*d*) The *code of the nervous system*. Although the amplitude of the nerve impulses in any one type of neurone may differ from that of another type, the amplitude is fixed for each individual class of nerves. The code of the nervous system is thus a frequency code rather than an amplitude code (*see* **15**).

6. Synaptic transmission. The junction between one nerve and another or between nerve and muscle is called a *synapse*. The arrival of the nerve impulse causes release of a chemical, usually *acetylcholine*, at the synapse, and this diffuses across the thin synaptic gap to the next nerve or muscle membrane. Here it sits at specific sites and causes a change in the membrane perme-

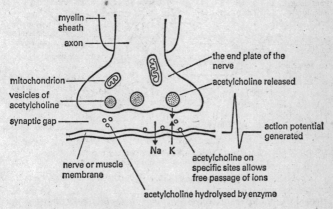

FIG. 109   *The synapse.*

The diagram shows the several events that occur at a typical synapse.

ability such that ions pass more readily than before. In turn this leads to the generation of an action potential. Meanwhile an enzyme, *choline-esterase* breaks down the acetylcholine and the original state is restored once more.

The type of synapse described above is excitatory, i.e. it leads to some other event occurring in the next nerve or the effector. Other synapses work in the reverse manner, that is to say chemicals produced at the synapse actually increase the membrane resistance to ion flow. Such synapses are inhibitory. Any given nerve will have both excitatory and inhibitory synapses in contact with it and whether or not the nerve will produce an action potential depends on the summation of both types of synaptic action.

7. Reflex pathways. That part of the nervous system which co-ordinates the activity of striated muscles and is under conscious controls is called the *somatic* while that which co-ordinates smooth muscles and glands and is not under conscious control is termed *visceral*. The visceral is also part of the autonomic system (*see* **8**).

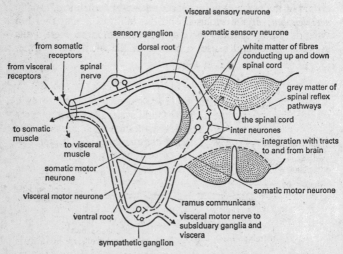

FIG. 110   *Transverse section through the spinal cord showing the somatic and visceral sensory and motor routes.*

This is a key diagram in understanding spinal reflexes.

Spinal reflexes involve both somatic and visceral nerves and the generalised pattern is shown in Fig. 110. It will be seen that the somatic motor neurones have their cell bodies contained within the spinal cord whereas those of the visceral arcs are found outside the cord in peripheral ganglia.

The neurones involved in the reflex paths are contained in the grey matter of the spinal cord whereas neurones involved in transmission up and down the cord are in the white matter.

Spinal reflexes are important in protective behaviour such as the withdrawal of the limbs from pain; they also play a vital role in locomotion and in standing. Initiation of complex muscular activity takes place in the brain and the impulses are transmitted down the spinal cord and pass out through the motor nerves to the muscles. Monitoring of the power generated by sets of muscles in a given situation is done by proprioceptors within the muscles operating through reflex pathways.

8. The autonomic system. This part of the nervous system, as already described, co-ordinates the activity of the smooth muscles and glands and operates below the level of consciousness. It comprises two largely antagonistic sub-systems, the *sympathetic* and the *parasympathetic* (*see* Fig. 111).

Some of the important characteristics and effects of these two parts of the autonomic system are seen in Table XIX.

PARTS OF THE BRAIN AND THEIR FUNCTIONS

9. The brain. The mammalian brain may be thought of as a massive development of the interneural system between receptor and effector systems. While there are interneurones between these systems in the spinal cord they are comparatively few in number. Information from the major receptors of the body, e.g., ears, eyes, nose, etc. is fed into the brain and there "computed" with the existing store of information to produce the initiation of appropriate impulses.

It is possible to localise specific functions in the various parts of the mammalian brain (*see* Fig. 112).

10. The hindbrain. The anterior part of the roof of the hindbrain is thin and permeable and makes up the posterior choroid plexus. Through this region nutrient substances enter the cerebro-spinal fluid of the fourth ventricle.

The medulla lies directly at the end of the spinal cord and acts

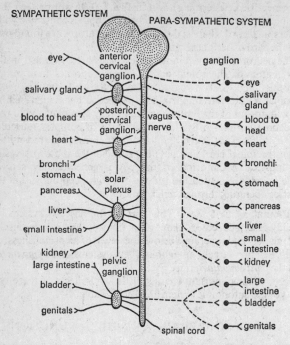

FIG. 111 *A simplified scheme showing the roles of the sympathetic and para-sympathetic systems.*

It should be noted that sympathetic neurones pass directly on to the organs they innervate while the parasympathetic neurones run by means of an intermediate ganglion.

partly as a main nerve trunk between the cord and anterior regions of the brain. Many important visceral activities have their regulation integrated within the medulla as for example heart beat and respiration. Most of the cranial nerves originate from the medulla.

Anterior and dorsal to the medulla is the cerebellum. This part of the hindbrain has a much convoluted surface and the internal arrangement of neurones mirrors the organisation of the cerebral cortex itself.

TABLE XIX. THE SYMPATHETIC AND PARASYMPATHETIC
NERVOUS SYSTEMS

	Sympathetic	Parasympathetic
Anatomical arrangement	The visceral components of the spinal reflex. The sympathetic ganglia are close to the spinal cord and there are long motor nerves which run directly to the organs concerned. In the mammal the segmental ganglia are coalesced into major units such as the solar plexus.	Consists of branches from the fifth and seventh and ninth cranial nerves, and branches from the sciatic region of the spine, though mainly however from cranial nerve ten, the vagus. The ganglia of the parasympathetic system are close to the organs concerned.
Synaptic transmitter	Preganglionic fibres acetylcholine. Post ganglionic fibres noradrenalin and related substances.	Preganglionic fibres acetylcholine. Post ganglionic fibres acetylcholine.
General effects	Increase of blood pressure by constriction of gut arterioles and increase of heart beat. Inhibition of peristalsis and digestive activity. Dilation of the bronchi. Secretion of sweat and erection of hairs. Dilation of pupil of eye. Contraction of spleen which increases blood volume. Release of glucose from liver. Secretion of adrenalin.	Decrease in blood pressure by slowing heart beat. Stimulation of peristalsis and digestive secretion. Constriction of bronchi.
Summary	The sympathetic system prepares the body to withstand stress situations. It works with the hormone adrenalin.	The para-sympathetic system is concerned with regulation of steady maintenance activities such as digestion. Its control is dominant in normal, i.e. non-stress, conditions.

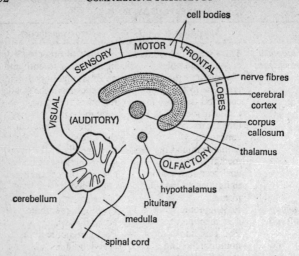

FIG. 112   *Generalised section of the human brain showing the main sensory areas.*

Into the cerebellum pass afferent nerves from the eyes and balance organs of the inner ear as well as from the proprioceptors of the body. There is also an input from the cerebral cortex. Integration of all this information leads to the regulation of posture, muscle tone and locomotion, the efferent fibres leaving the cerebellum and passing, via the medulla, to the spinal cord.

11. The midbrain. The optic lobes that make up the roof of the midbrain in the lower vertebrates are vestigial in the mammal and the visual input is directed forwards to the cerebral cortex. The movement of the eyes is controlled by the roof of the midbrain.

The floor of the midbrain is made up of the *cura cerebri* which connect together the fore and hind brains. This region also contain the red nucleus which is of special importance in the integration of cerebral and cerebellum activity in posture and locomotion.

12. The forebrain. In the primitive vertebrates the forebrain is largely concerned with olfaction but in the higher forms it is enormously expanded and has taken over many other functions, acting as the dominant region of the brain.

The forebrain consists of the following parts:

(a) *The thalamus.* This is in the posterior part of the forebrain and it acts as a junction box between the various parts of the former and the other regions of the brain. It also integrates activity of the different functional regions of the forebrain itself.

Through the thalamus pass both efferent tracts from the body and afferent motor pathways.

(b) *The hypothalamus.* This lies below the thalamus and is the control centre of many visceral activities, such as the control of temperature and osmotic regulation. It may also be the site of initiation of quite elaborate aggressive or defensive behaviour patterns.

The hypothalamus is situated near the pituitary, the main endocrine organ of the body and it works together with this gland. Some of the secretions of the pituitary such as ADH are synthesised in the hypothalamus and transported to the pituitary.

(c) *The cerebral cortex.* The roof of the forebrain is very extensive in mammals and various functional regions have been described by biologists. The posterior part is the visual cortex where optic stimuli are received and interpreted. Laterally is the auditory area for the reception of sound. In front of the visual cortex is the somatic sensory area where the parts of the body are "proportionately represented" according to the number of touch sense organs present (i.e. many for the hands and tongue, few for the back). The anterior ventral region is concerned with olfactory processes.

Between the somatic sensory and the frontal lobes lies the somatic motor area. Here motor impulses which control the whole movement activity of the body are initiated although, as explained above, they integrate via the red nucleus with the output of the cerebellum.

Each part of the above sensory region is represented according to the habits of the mammal concerned. Arboreal animals such as monkeys have large visual and auditory areas and small olfactory regions, whereas insectivores and carnivores have very large olfactory lobes.

The frontal lobes are not well developed in mammals other than man where they are very conspicuous. These are the so-called "silent areas" of the brain and appear to be the seat of intelligence and personality and the various other behavioural differences that exist between man and other mammals.

SENSORY RECEPTORS

13. The role of sense organs. The mammal lives in an environment in which all sorts of changes are taking place, both around it as well as within its own body.

(a) *External changes.* Those external changes of biological importance are variations in the wave length and intensity of light, variations in molecular shapes which are detected as smells or tastes, and mechanical or pressure changes detected as hearing or touch.

(b) *Internal changes.* It is important that the body monitors the changes that are taking place within it so that homeostatic adjustment can be made. Chemical and osmotic changes, temperature fluctuations, mechanical changes and gas tensions are some of the important pieces of information that are fed into the central nervous system by the internal receptors.

14. Receptor cells. While it is still far from clear how receptors work, some general aspects of their functioning have been established. All receptors cause the generation of impulses either from themselves or from the cells immediately in contact with them, e.g. the rods and cones of the eye. It seems that the particular stimulus that the receptor is specialised to detect causes a depolarisation of the cell membrane and the initiation of an action potential.

In many sense cells, e.g. touch receptors, prolonged stimulation leads to adaptation so that impulses are no longer generated (clearly in pain receptors this does not occur).

The code of the receptors, and of the nervous system as a whole, is a frequency rather than an amplitude code. Intense stimulation thus causes the receptor to produce impulses at high frequency but not to produce higher amplitude impulses. How this works in communication is seen in Table XX.

In the above case shown in Table XX the receptors work at different thresholds and over a certain range of stimulation. In this sort of way very complex degrees of stimulus discrimination can be assessed by the body.

15. Resting discharge. So far it has been assumed that the unstimulated receptor does not discharge until it is affected in some way. This is not always the case as the mechanical receptors of the muscles and semi-circular canals tend to produce a *resting discharge* without stimulation.

TABLE XX.   INTENSITY OF STIMULATION AND ACTIVITY OF
RECEPTOR

Stimulus	Receptor	Discharge per second
Light touch	Light pressure receptor	2
increased	Light pressure receptor	20
Medium touch	Medium pressure receptor	2
increased	Medium pressure receptor	20
Heavy touch	Heavy touch receptor	2
increased	Heavy touch receptor	20
Pain	Pain receptor	2
increased	Pain receptor	20

The advantage of a resting discharge is that its frequency can be altered in either direction, i.e. it can be increased or abolished. In the cases mentioned the constant volley of impulses from the muscles and inner ear seem important in the maintenance of the whole muscle tone throughout the body.

16. The classification of sense organs. Receptors may either be classified in terms of the type of stimuli they detect or else by whether they respond to the external or internal environment.

(a) *Classification by stimulus:*

(i) Chemoreceptors: for smell and taste, pH, osmotic concentrations, gas concentrations.

(ii) Mechanoreceptors: for sound, pressures, touch, tension of muscles and tendons, vibration.

(iii) Photoreceptors: for light, normally the visible spectrum but some receptors may detect further into the ultra-violet and infra-red regions than our own.

(iv) Thermoreceptors: for temperature, e.g. of skin or of blood.

(b) *Classification by environment:*

(i) Exteroreceptors: for changes in the external environment, e.g. teloreceptors for ear or eye for change at a distance, cutaneous receptors for changes at skin surface, and chemical receptors for taste and smell.

(*ii*) Interoreceptors: for changes within the body, e.g. proprioceptors are within the muscles, visceroreceptors are those in the viscera and chemical interoceptors for chemical changes in the body.

17. The eye. (A basic knowledge of the structure and function of this organ are assumed from "O" Level studies. Here are included some more detailed points more suitable for "A" Level answers on this topic.)

(*a*) *Accommodation.* This is the formation of a sharp image on the retina. It is achieved partly by the refraction due to the cornea which has a refractive index of 1·34 which is equivalent to a power of 40 diopteres (1 dioptere = 100/focal length in cm). The cornea is responsible for about 70 per cent of the refraction of light entering the eye.

The role of the lens is subservient but critical in obtaining a sharp focus. It is held under tension by the ciliary ligaments and at rest has a long focal length suitable for distant objects. When near objects are viewed the ciliary muscles contract taking the strain off the lens whose radius of curvature increases. This in turn shortens the focal length. This action of focussing is involuntary.

(*b*) *Aberrations.* Spherical aberration is minimised by the special curvature of the lens which corrects for this defect. Chromatic aberration which is due to the differences in refraction of different wavelengths is reduced by having a slightly yellow lens and by a more distinct yellow spot on the retina. The yellow takes up the blue wavelengths which show most serious chromatic aberration. The retina itself is also some ten times more sensitive to orange/red wavelengths which show least aberration.

(*c*) *Role of the iris.* The expansion and contractions of the iris diaphragm are brought about by two sets of antagonistic muscles, the radial opening the pupil and the circular muscles contracting it. The response is automatic and triggered by receptors within the iris itself; it may also be brought about by adrenalin. The function of the iris is to stop down incoming light to obtain an optimum illumination at the surface of the retina.

(*d*) *Events taking place at the retina.*

(*i*) *The rods* are sensitive to light intensity below that of the cone cell receptors. There are some 120×10^6 in the human retina and they detect "black and white" (as we shall see below it is also possible that the rods are responsible also for detecting the blue end of the spectrum). The structure of a rod is shown in Fig. 113.

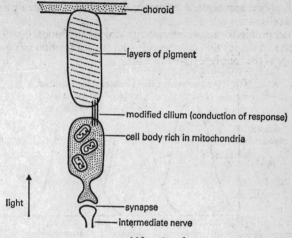

FIG. 113 *A rod.*

Within the pigment containing end of the cell the visual purple or rhodopsin is broken down by the light, with the release of chemical energy. This in turn, via mechanisms in the rest of the cell, leads to the formation of an action potential.

$$\text{Rhodopsin} \underset{\text{dark}}{\overset{\text{light}}{\rightleftharpoons}} \text{Retinene} + \text{Opsin} + \text{Chemical energy}$$
(visual purple)

There may be as many as 150 rods sharing the same final nerve path into the back of the cerebral cortex where the "seeing" actually takes place. It will be noted that the retina is inverted so that the light has to pass through nervous tissue before reaching the sensitive part of the receptors.

(*ii*) *The cones.* In the human eye these are concentrated at the yellow spot, or fovea, where the incoming light automatically focusses. Here they are found at a density of some $1,000/mm^2$, and the overlying neurones are parted so that there is less cytoplasm between the light and receptors. Cones are only stimulated by high intensity light, which is why we cannot see colours in dim light. It is probable that there are several types of cones, some having a pigment erythrolabe with an absorption peak in the red band, others containing chlorolabe which takes up green light. A cone containing cyanolabe, i.e. absorbing blue light, has still only a hypothetical existence; it may be represented by the rod cells.

According to the wavelength arriving at the retina it will be seen as a colour that corresponds with the stimulation of one or more of the cone types (*see* Fig. 114).

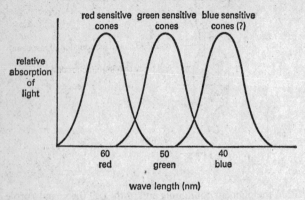

FIG. 114 *Absorption spectra of various cones.*

Cones have a one to one nerve connection from the retina to the cerebral cortex.

(*e*) *Why all the intermediate neurones and ganglia?* The answer to this is not certain but recent investigations show that the great amount of nerve tissue in the retina may serve to unscramble incoming information and very rapidly reduce it to its essentials. This would be very important in survival where an animal has to act on incoming image changes very rapidly. All the nerve tissue leaves the retina at the *blind spot* and here there are no receptors so light falling on the blind spot is not perceived.

(*f*) *Distance judgment.* In animals such as man, with an arboreal ancestry, the eyes are set apart and both eyes see very much the same field of view. In fact the eye muscles behind the sclera of the two eyes are linked together and differences in their tensions to obtain an exact focus are interpreted as distances of the object from the eyes. This is called stereoscopic vision and is essential in climbing animals and in carnivores. Mammalian herbivores, such as horses or rabbits, tend to have a much wider field of view but cannot judge distances with the same precision as those with stereoscopic vision.

18. The ear. (A basic knowledge of the structure and function of the ear will be assumed from "O" Level studies. Some details of ear mechanism more appropriate for "A" Level have been included below.)

(a) *Functions related to balance and posture.* The superior part of the inner ear consists of the sacculus and utriculus, together with the three semicircular canals. Otoliths are found in the first two chambers and these sit lightly attached above a group of neuromasts, nerve cells sensitive to mechanical deformation.

If the body is subjected to linear accelerations, that is acceleration forwards, backwards, sideways or up and down, the otoliths hang back at the start of such movements and push forwards as they cease. It is at the starting and stopping of the acceleration that the movement is sensed as the neuromasts discharge via the VIIIth (auditory) nerve to the co-ordination centre in the cerebellum.

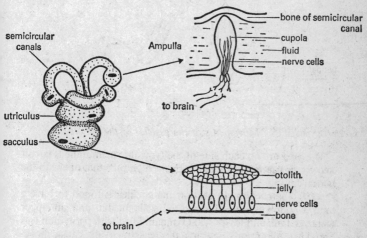

FIG. 115 *Structure of the inner ear organs of balance and posture.*

Angular accelerations, whereby the body is rotated in any one of the three planes of space, are detected by the ampullae of the semicircular canals. The gelatinous cupola in which the ends of neuromast cells are embedded hangs back at the start of rotation and stimulates them to change their rate of discharge. Deceleration at the end of the rotation has the reverse effect, so once again it is the change of movement that is detected.

From the otoliths and from the ampullae a steady stream of information pours into the cerebellum which uses this, together with other data from the eyes and the muscle receptors, to control posture and balance.

(b) *Functions related to hearing.*

(i) *The route into the inner ear.* The pinna, with its trumpet shape, collects incident sound waves which are directed down the external ear to the tympanic membrane. This has an area 22 times greater than the fenestra ovalis and by means of the three ear ossicles the force at the membrane is magnified by this amount. The ossicles have a very low mass, of some 3 mg total, and they move about each other as indicated in Fig. 116.

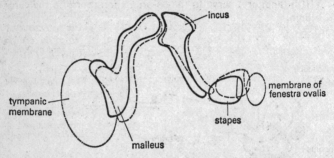

FIG. 116 *Movements of the ossicles of the middle ear.*

The pressure on each side of the tympanic membrane is equalised by air passing through the Eustachian tube that opens to the buccal cavity.

(ii) *The cochlea.* In section the cochlea is found to have a central endolymph-filled tube, the scala media, and an upper scala vestibuli and lower scala tympani filled with perilymph (*see* Fig. 117). Vibrations entering the fenestra ovalis cause movements of the basilar membrane. On this is sited the Organ of Corti where mechanical energy becomes converted to nerve impulses.

(iii) *The Organ of Corti.* The organ is so built that lateral forces result from the up and down vibrations of the basilar membrane. The effect is very subtle, but the movements are increased in amplitude by the arrangement of rigid and flexible

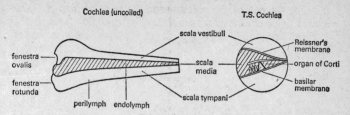

FIG. 117   *The cochlea.*

parts of the organ. The system is so extremely sensitive that movements no greater than 10^{-9} cm produce a nerve impulse.

The neuromasts have fibres of different lengths and the short ones respond to higher frequencies than the long ones. There are some 20,000 such neuromasts in the human cochlea with those responding to high frequencies (up to 16 kH) at the top end and the low ones at the bottom end. Much of the length of the coclea is taken up by nerve cells that respond at the frequencies of the human voice (*see* Fig. 118).

The nerve impulses generated by the Organ of Corti pass via the auditory nerve to the auditory region of the cerebral cortex.

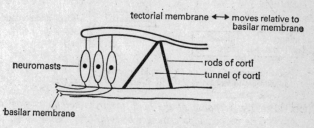

FIG. 118   *The organ of Corti.*

MAMMALIAN ENDOCRINES

19. The nature of endocrines. A brief introduction to the nature of endocrines was given in **1**. It will be recalled that they work with the nervous system in the general co-ordination of the body but, on the whole, they control various long-term changes such as maturation and growth.

The chief endocrine organ is the *pituitary* and its secretions tend

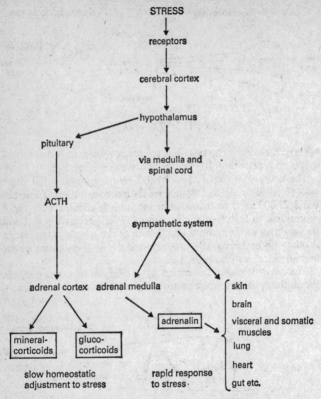

FIG. 119 *The response of the body to stress situations such as injury, danger or cold.*

Adrenalin fortifies the action of the sympathetic nervous system which prepares the body to withstand stress. The action of the cortical hormones is slower but very important in long-term recovery or adjustment.

to affect all other endocrine glands. There may also be a further hierarchy within the latter themselves.

NOTE: For the purposes of rapid assimilation the information on endocrines required by students will be given in the form of a table (*see* Table XXI) and annotated diagrams. These, between them, contain the essential information required (*see* Figs. 119 and 120).

20. The action of hormones at the level of the cell. A very active area of contemporary research in biology is the way in which hormones operate at the level of the cell and the gene. Hormones

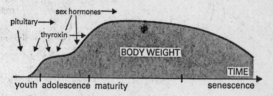

FIG. 120 *The importance of the major hormones in the life cycle of a mammal.*

Growth is directly controlled by a pituitary hormone as well as by thyroxin. Both these hormones are important in the initiation of sexual maturity and the sex hormones themselves bring about changes associated with maturation.

may be simple molecules derived from single amino-acids, as is the case with thyroxin and adrenalin, or they may be polypeptides such as insulin or again they may be steroids, based on the cholestrol molecule, as is the case of the sex and corticoid hormones. All hormones act in minute concentrations, say less than 1 p.p.m., and all tend to effect specific target cells, often bringing about a range of effects after a certain time lag. All these phenomena are on the way to an explanation in these recent findings.

(*a*) *Hormones and genes.* It is known that all the cells in the body except the sex cells have the same complement of genes, yet some cells differentiate in one way and some in another; some make certain proteins and enzymes, whereas others produce quite different kinds. The relationship that exists between some hormones and genes is now more fully understood.

(*i*) *Ecdysone and chromosome puffs.* One of the classical studies on the relationship between hormones and genes was made on the larvae of midges. Here the injection of the metamorphosis hormone ecdysone causes, after a few minutes, enlargements or puffs to appear on certain regions of their chromosomes. Experiments using labelled bases and autoradiography indicated that mRNA was being actively synthesised at such

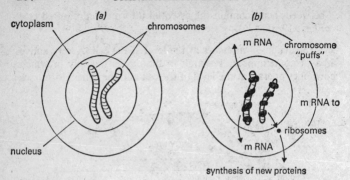

FIG. 121　*Effect of ecdysone on the cells of midge larvae.*

(*a*) Resting cell nucleus. (*b*) Cell after treatment with ecdysone.

sites. This in turn would be associated with protein manufacture in the cytoplasm shortly afterwards (*see* Fig. 121).

The implications were clearly that the hormone was "switching on" those parts of the chromosomes where genes were involved in the manufacture of mRNA encoding specific proteins involved in metamorphosis.

(*ii*) *Hormones as activators of genes.* Clearly the majority of genes in any given cell are inactive, possibly masked by inert combination with the ubiquitous protein *histone*. Some hormones appear to activate specific gene areas, either directly, or else by combining with repressor molecules (*see* V, **23**) and to lead to rapid synthesis of mRNA as well as other types. These latter activities are associated with subsequent protein synthesis in the cytoplasm.

It is clear that some cells, so called *targets*, are much better able to translate the instructions of given mRNA than are others. Thus mRNA encoding information for haemoglobin synthesis works much more rapidly in red blood corpuscles than muscle cells. Presumably this is because the former have many more active ribosomes for haemoglobin synthesis than the latter, which in turn might be much faster at myosin synthesis.

The effect of several different hormones on rat liver cells is shown in Fig. 122.

Once again the inference must be that the hormones are acting directly on the DNA in the nucleus, that is on specific genes, and

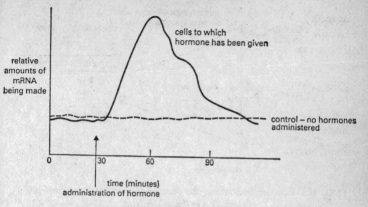

FIG. 122 *Effect of insulin, thyroxin or testosterone on rat liver cells.*

thus leading to increases in mRNA and tRNA molecules. The hormones that are known to effect RNA synthesis are as follows:

> pituitary growth hormone;
> thyroxin;
> adreno-cortico-tropic hormone;
> insulin;
> sex hormones;
> some corticoid hormones.

(*b*) *Hormones and cell activities not involving genes.* Not all hormones activate genes. Some act on enzyme systems already present in their target cells. It is known that insulin, glucagon and adrenalin all act in this way and it seems very likely that a second messenger is involved on the target cell surface.

The hormone molecules react with specific receptor molecules at the cell surface and this in turn leads to the release of another intermediate chemical which activates enzyme systems within the cell. Thus insulin leads to stimulation of the hexokinase enzyme, which leads to phosphorylation of glucose that has entered the cell. This allows the glucose to enter into other metabolic reactions and more glucose will be taken up across the cell membranes as the level inside falls.

Adrenalin, on the other hand, promotes glycogen break down through activation of cyclic AMP (adenyl monophosphate)

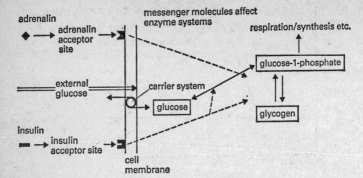

FIG. 123 *Scheme indicating the possible action of adrenalin and insulin on cell metabolism.*

Adrenalin promotes glucose release, insulin its uptake.

enzyme systems. These AMP systems are important in the initial stages of cellular respiration pathways (*see* Fig. 123).

Hormones are inactivated either at the cell surface or within the cells on which they act. By feedback monitoring from the original endocrine gland the levels of hormones within the body are kept fairly constant. Application of a hormone (as for example the progesterone contraceptive pill) suppresses the production of the particular endocrine by the body.

THE NERVOUS SYSTEMS OF *PARAMECIUM* AND THE EARTHWORM

21. Paramecium. This is a ciliate protozoan and as such it has the most elaborate system of organelles found in single-celled animals. However it now appears that the neuromotor system described for this organism is the result of a staining artefact. It is not seen under the electron microscope.

The fibrils that link up cilia have been described as the basis of co-ordination of their beat but this has not been proved. When the organism encounters an unfavourable stimulus or an obstacle the ciliary beat is reversed. Paramecium then reorientates and advances again. This trial and error behaviour leads to aggregation in favourable conditions.

In harmful situations, such as attack by a *Suctorian*, the paramecium discharges trichocysts.

22. The earthworm: general organisation of the nervous system.

(a) The cerebral ganglia are paired structures lying dorsal to the gut in segment 3, and from them sensory nerves pass into the prostomium which is well supplied with touch, light and chemical receptors (see Fig. 124).

(b) Two circumpharyngeal commissures run round the gut and join the ventral nerve cord which has a ganglion in each segment (unlike insects not all the cell bodies are located in the ganglia). At the point of junction of the commissures and the ventral nerve cord is the sub-pharyngeal ganglion.

Nerves from the commissures supply the peristomial region while others from the sub-pharyngeal ganglion supply segments 2, 3 and 4. Part of the autonomic supply to the viscera also originates from the circumpharyngeal commissures.

(c) The ventral nerve cord supplies each segment with three pairs of nerves which include both efferent and afferent pathways. There is also a segmental component of the autonomic system which is antagonistic to the commissure fibres.

Sense organs as described for the prostomium are present all over the body surface and there are also proprioceptors within the muscles.

23. Earthworm: the giant fibres.
An important feature of the earthworm's nervous system is the presence of giant fibres in the dorsal part of the nerve cord.

There are two lateral and one median giant fibres and these fibres are able to conduct impulses at some 12 metres per second rather than the 0·025 metres per second operating in the rest of the nervous system. Sensory input goes into the median giant from the front end of the worm and the fibre conducts from the head end to longitudinal muscles throughout the body. Sensory input enters the lateral giant fibres from the posterior end of the worm and they also connect segmentally with longitudinal muscles.

Sudden stimuli reacting on the giant fibres allows the earthworm to contract very rapidly and is of importance in its survival behaviour.

24. Earthworm: reflexes in locomotion.
Locomotion is normally initiated from the cerebral region and waves of alternate contraction of longitudinal and circular muscles pass down the worm. There is also an inter-segmental reflex pathway whereby proprioceptors in the muscles of one segment are connected to the

TABLE XXI.   THE ENDOCRINE

Organ	Origin and Structure	Secretion
Pituitary	It is situated below the hind part of the forebrain, below the hypothalamus. The anterior part is derived from an up-growth of the buccal cavity called the hypophysis and the posterior part from a down-growth of the brain called the infundibulum. The pituitary is close to the teloreceptors and brain and is the "master" endocrine gland of the body. It produces a large number of both general and specific hormones.	Pituitrin, the growth hormone. (somatotrophin) Thyrotropic hormone. Adrenocorticotropic hormone. Pancreotropic hormone. ?Diabetogenic hormone. ?Ketogenic hormone. Gonadotropic hormones. (i) F.S.H. (ii) I.S.C.H. (iii) Lutenising hormone. (iv) Prolactin. Oxytocin. Vasopressin. Antidiuretic hormone.
Thyroid	It is a lobed organ in the region of the larynx. The large glandular cells secrete a colloid which contains the hormone. There are four iodine atoms in the molecule of hormone.	Thyroxin+tri-iodo-thyronine.
Parathyroids	These are small lobes in the posterior part of the thyroid.	Parathormone.
Pancreas	This originated in primitive chordates from cells in the wall of the gut; in more advanced forms it is embedded with secretory cells in the pancreas.	Insulin. Glucagon
Adrenals	In lower vertebrates the two parts of the adrenal are separate and come respectively from sympathetic cells and those near blood vessels of the gonads. In higher vertebrates there is a composite gland which has an outer cortex producing a variety of hormones and in inner medulla producing adrenalin.	Various cortical hormones. Adrenalin.
Gonads	The nature of the gonads is initially determined by the genetic constitution of the individual. Endocrines are synthesised from both the cells involved in gamete manufacture and the interstitial cells of the gonad.	Female oestrogen and Progesterones. Male testosterone.

GLANDS AND THEIR HORMONES

Effects of Hormone	Results of Hormone excess and deficiency
Encourages growth by promoting protein synthesis, especially important in early development.	*Excess:* In early growth "giantism" is produced with abnormal but proportional size. Later in development excess of the hormone leads to enlargement of the limb extremities and jaw where bone growth is not complete. This condition is called agromegaly.
Stimulates activity of thyroid. Stimulates adrenal cortex activity Stimulates pancreatic activity. It is antagonistic to insulin. This causes production of ketones.	*Deficiency:* Produces "dwarfism" where a very small individual is produced. Provided thyroid function is normal such a dwarf would be in all other respects normal.
Initiates egg and sperm development. Stimulates interstitial cells of gonads to secrete sex hormones. Stimulates corpus leteum development.	Causes disturbances of normal function of these glands.
Causes contraction of the uterus. Increases the blood pressure. Controls water content of the urine.	
Stimulates the gonads to produce sex hormones and regulates the metabolism of the body.	*Excess:* Causes gross increase in the metabolic rate and a condition known as "Graves disease". *Deficiency:* A congenital deficiency leads to cretinism which is the failure to mature sexually, physically or mentally. In adults there is a general slowing down of the metabolism.
Causes the release of phosphate ions and loss in the urine leads to disturbance in the calcium equilibrium between tissues and bone.	*Excess:* Causes withdrawal of minerals from the bones. *Deficiency:* Causes a lowering of the blood and tissue levels of calcium which leads to tetany (rigid locking of muscle).
Allows the uptake and utilisation of sugars thus reducing its level in the blood. Encourages conversion of sugar to tissue glycogen and fat and amino-acids. Glucagon reverses effects.	*Excess:* Very low level of blood sugar prevents normal interchange with active tissues and failure of respiratory functions. *Deficiency:* Leads to a high level of blood sugar and a variety of osmotic and metabolic disturbances. The disease resulting from these upsets is diabetes mellitus.
A variety of hormones which help regulate the ionic and sugar balance of the body as well as its ability to respond to stress.	*Excess:* Causes abnormal levels of blood sugars from endogenous breakdown of tissues and muscular weakness. *Deficiency:* Leads to aggravated response to stress as seen in Addison's disease.
The hormone is a synaptic transmitter for the sympathetic nervous system and leads to preparation of this system to combat stress conditions.	There is little material available here but it is known that removal of the adrenal medulla in experimental animals causes death to occur under very normal stress conditions.
The effects of these hormones are described in relation to sexual co-ordination in X, 16.	*Excess:* Effects are difficult to analyse but probably lead to extreme personality effects in the male or female concerned. *Deficiency:* Leads to sterility. It should be noted that sex hormones are also regulated via the thyroid and adrenal cortex endocrines.

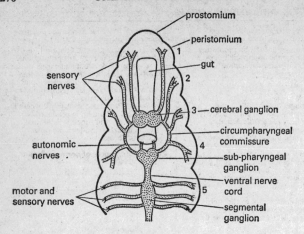

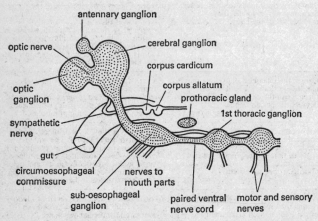

FIG. 124 *Cerebral regions of the earthworm and the insect nervous system.*

The degree of cephalisation is much greater in the latter animals.

muscles in the next segment. When the muscles in the first segment contract impulses pass to the second segment and a wave of contraction results.

When the longitudinal muscles are contracted (making the

segment short and fat) the chaetae muscles are also contracted so that firm contact is made with the ground.

Rhythms of electrical activity corresponding to the frequency of peristalsis have been detected in the ventral nerve cord.

25. Earthworm: transmitter substances. The presence of acetylcholine and adrenalin has been reported for the earthworm and synaptic transmission is doubtless similar to that of mammals.

SOME ASPECTS OF CO-ORDINATION IN INSECTS

26. The nervous and endocrinal co-ordination of insects. This is of considerable interest as this group is by far the most successful of all animals in terms of the number of species, and the development of complex means of co-ordination is at least one of the major reasons for this success.

27. The brain. While the general pattern of the nervous system conforms to the type already described for the annelids (insects being derived from and closely related to annelids) there are certain major developments (*see* Fig. 124).

The mass of the brain is concerned with the reception of input from the major sense organs in the head. It contains large numbers of association neurones and integrates data from incoming stimuli. Concerted activity of the whole body in response to these stimuli is initiated by the brain but carried out as a direct result of motor ganglia in the segmental cord.

Removal of the brain of an insect grossly disrupts normal function and activity becomes exaggerated and repetitive.

The sub-oesophageal ganglion contains motor centres and controls the movement of the feeding appendages. A small frontal ganglion is associated with autonomic pathways.

28. The thoracic ganglia. There are three of these, one for each segment of the thorax in most insects, although in some groups such as flies they are fused into one large mass.

The main function of the thoracic ganglia is the control of locomotory activities such as flying and walking. Proprioceptors in the legs and wing muscles (or haltares in flies) feed information via sensory neurones to the appropriate thoracic ganglion and motor nerves bring back the impulses that lead to co-ordinated movement.

There is a hierarchy within the thoracic ganglia from anterior to posterior so that the whole activity of this region operates smoothly.

All the ganglia of the nerve cord have the means of controlling respiratory activity of the particular segments in which they are situated.

29. The abdominal ganglia. There are usually six of these with the last being enlarged and representing a number of fused ganglia. The abdominal ganglia, besides the work of regulating the spiracle activity of their own segments, co-ordinate the visceral and reproductive functions associated with the abdominal region.

30. Giant fibres. Some insects that are capable of very rapid response, such as cockroaches, have giant fibres which conduct stimuli from the abdomen of the insect, e.g. the anal cerci, forwards to the brain. The speed of conduction of these fibres is comparable with the medullated fibres of mammals and they are used for escape reactions.

NOTE: The details of insect sense organs will not be described here (*see* the Bibliography for further reading). It is sufficient to appreciate that they are very much more complex than those of most invertebrates and relay to the insect a great deal of detailed information about changes taking place in its environment. In many cases, and in particular with chemo- and photo-receptors, insects may possess sense organs that are one hundred or even one thousand times more sensitive than those of man, e.g. the ability of certain male moths to scent a female over many hundred metres. There is no doubt that the success of insects is partly due to the complexity of their behaviour and this in turn can only be because of the elaborate development of sense organs and data analysis by the brain.

31. Endocrine co-ordination in insects. As with the vertebrates, hormones or endocrines of insects co-ordinate such long term activities as reproduction, growth, moulting and metabolic rates. As moulting and metamorphosis do not occur in the majority of vertebrates a short description of the role of insect hormones in these processes will be given (*see* Fig. 125).

In those insects where there is incomplete metamorphosis, such as the cockroach, the supply of juvenile hormone gradually declines and each moult brings the nymph nearer to the adult form. In those, such as the butterfly, where there is a complete metamorphosis the supply of juvenile hormone is very suddenly

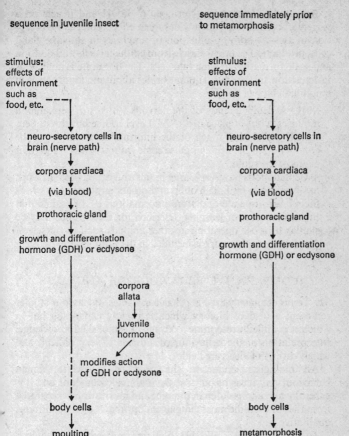

FIG. 125 *The activity of insect hormones involved in growth and moulting.*

cut off and very pronounced changes occur during a pupal stage.

It has been possible to produce various insect "monsters", i.e. very large immature forms or very tiny adults by transplantation of the appropriate endocrine gland.

The way in which ecdysone induces RNA synthesis in certain cells has been described in **20**. This hormone is the same as the growth and differentiation hormone or GDH and it acts on a variety of cells causing such widespread effects such as reabsorption of the fat body, production of amylases in salivary gland cells, production of pupal cuticle from epithelial cells and division and scale formation in wing bud cells. The hormone is analagous in its action to the thyroxin that brings about metamorphosis in amphibian larvae.

NOTE: A detailed study of the nervous systems of cephalopod molluscs is out of place in an "A" Level course but the student should realise that much of the information we have on the conduction of the nerve impulse came from studies on the giant axons of the squid.

Cephalopods have very large brains and a great deal of work has been done by J. Z. Young and others on the way in which their brains operate and store information. It is possible that the mechanisms of learning described for the octopus may be similar to those operating in other animals such as ourselves. A reference to this work is included in the Bibliography.

COMPARATIVE BEHAVIOUR OF ANIMALS

32. Terms commonly used in behaviour studies. Behaviour studies represent a field of biology which is rapidly expanding but in which it is difficult to outline a corpus of established knowledge. Interpretation of the behaviour of animals is very difficult and widely divergent views are held.

All biologists emphasise the need to dissociate human behaviour and ideas of purpose from animal behaviour and it is generally agreed that human thought and motivation is separated from that of even the most "intelligent" primates by an enormous qualitative gap.

To some extent it is possible to relate complexity of behaviour to complexity of the nervous system but this is not always the case. Thus the reptiles with a substantial neopallium in the cerebrum seem to show no more complex behaviour than fish with no neopallium. The protozoan *Stentor* shows at least four types of response to unpleasant stimuli in a graded sequence and yet it has the simplest of neuromotor organelles.

Terms that are useful in behaviour studies are as follows:

(*a*) *Taxis:* a movement of the whole organism towards or away from a directional stimulus.

(b) *Kinesis:* change in the rate of movement or in the direction of movement in response to a simulus.

(c) *Reflex:* stereotyped response of a part of the body to a given stimulus.

(d) *Instinct:* an innate behaviour pattern of some complexity set off by an initial stimulus and normally depending on the hormonal state of the organism concerned.

(e) *Learning:* the modifying of behaviour in response to a given stimulus.

(f) *Intelligence:* the ability to reason, to assess a given situation and to work out the effects of any course of action in the mind.

33. Trends in the evolution of complex behaviour. Among the lower animals we find behaviour patterns are stereotyped and to a large extent mechanical. In certain higher groups such as birds and insects there are elaborate instincts developed and in the former some degree of modification of these instincts is possible making the response more adaptative to the situation. In the highest mammals there are glimmerings of intelligence or reasoning and this is very highly developed in man.

While it is difficult to generalise on this topic something of the main trends in the various types of behaviour is seen in Fig. 126.

34. Simplest types of behaviour. Taxis and kinesis involving the whole body and reflexes involving only a part of the body make up most of the behaviour of lower animals. The former types of response enable the organism to find and remain in favourable conditions and to avoid or to leave unfavourable ones and the latter are nearly always protective or involved in co-ordination of movements.

Some examples of simple behaviourisms are as follows:

(a) *Taxis:*

(i) +phototaxis (towards light): *Euglena* and other phytoflagellates.

(ii) −phototaxis (away from light): earthworm, woodlouse, *Planaria*, blowfly larva, *Amoeba*.

(iii) +chemotaxis (towards a chemical, i.e. up its gradient): sperms, *Paramecium*, *Planaria*.

(iv) +rheotaxis (against a current of water): most freshwater animals, e.g. *Planaria*.

(v) −geotaxis (against gravity): cercaria larva of flukes, glochidium larva of swan mussel.

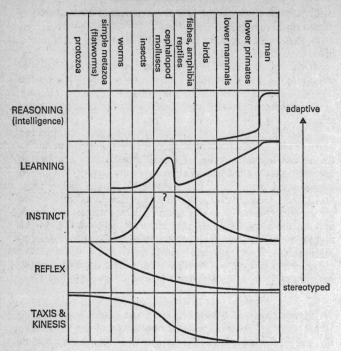

FIG. 126 *A very simplified scheme indicating the extent to which various types of behaviour are developed in different groups of animals.*

(*b*) *Kinesis:*

   (*i*) hydrokinesis (increased rate of change of direction in damp): woodlouse.

   (*ii*) chemokinesis (increased rate of change of direction in presence of a particular chemical): *Planaria*.

(*c*) *Typical reflexes* are as follows:

   (*i*) Crossed extension reflex of vertebrate limbs.

   (*ii*) Toe curling in primates.

   (*iii*) Pupil contraction.

   (*iv*) Withdrawal of limb from painful stimulus.

The mechanism of these simple types of behaviour whether it

is differential excitability in the sides of the nervous system or spinal pathways in vertebrates is fairly well established.

35. More complex forms of "fixed" behaviour. Instincts are more elaborate behaviour patterns. They are found to be highly developed in insects, birds and many mammals.

Some examples are the following:

(*a*) Migration, e.g. birds, certain crustaceans, mammals and some insects.

(*b*) Reproduction, e.g. courtship and display ceremonies in birds and many insects, feeding responses of parent and offspring as in nestlings, territory behaviour and aggression to rivals' nest building, etc.

(*c*) Feeding, e.g. frogs attack moving small objects, pecking behaviour of chicks, selection of hosts by parasitic flies.

NOTE: In the majority of cases these sort of reflexes are very exactly determined as can be shown by human interference producing what is in effect a "ridiculous" result, i.e. one that is non-adaptive. Thus a spider whose web's radial strands are cut will continue to try and complete the circular ones although the shape of the web is ruined. Mason bees will seal an unprovisioned cell if the food has been removed: ants can be made to march in circular column by joining the head of a column to the tail: sticklebacks and robins will attack red objects of cardboard or wool as if they were males of the species: sticklebacks will also do courtship behaviour to brown painted cardboard and gulls will try to incubate golf balls while gull nestlings will gape at yellow and red painted sticks.

Such phenomena indicate the mechanical nature of instinct: in no sense can the animal be thought of as "knowing what it's doing" nor indeed of having any idea of the object of its behaviour.

Of course in nature such interference does not occur and the particular instinct if triggered off by the right stimulus and at the right time will bring about a result (i.e. nest building, mating, feeding, etc.) which is of survival importance. Instincts seem to take up less nerve cells than more adaptable behaviour and they can be elicited in full complexity without learning from another member of the species. Clearly, instinctive behaviour is very highly appropriate to the lives of most animals.

36. Learning. There are various degrees of learning, and although this process is important to intelligent animals, it is also found associated with those whose behaviour is largely instinctive. For example, recently it has been shown that chaffinches and other birds kept isolated from their own species are not able to sing as well as other birds.

(a) *Imprint learning.* This is important in many young birds such as ducks where a "mother" figure is imprinted on the nestling soon after hatching. Here again it is possible to imprint a substitute mother such as a plastic toy duck on the young bird. The same behaviour is found in lambs.

(b) *Trial and error learning.* This is much employed on animal behaviour but also occurs in nature, as for example the learning to leave alone distasteful or poisonous insects. In the laboratory the animal may be rewarded for a correct move or punished for an incorrect one and in this way mazes and puzzles can be learned.

(c) *Habituation.* This is a further form of learning where a repetitive stimulus eventually becomes ignored. Classical conditioning as established by Pavlov depends on training the animal to respond in a normal reflex manner to an abnormal stimulus, e.g. the ringing of a bell causing salivation in a dog.

(d) *The "learning curve".* In most experiments on learning, improvement in the performance of a particular action is noted with repetition. It is not possible to say whether the learning curve represents a gradual learning process or whether there is a point between one trial and the next when the individual suddenly "gets the idea".

Learning is of great importance to all young mammals who depend on copying their parents for the development of their own survival-behaviour patterns. Clearly learning gives a greater adaptability to a changing environment than does rigid instinct patterns.

37. Intelligence or reasoning. As stated above this type of behaviour is only found in a few mammals such as man and closely-related species. This type of behaviour is very efficient as it is highly adaptable to a given situation and tends to prevent the animal making mistakes in response.

In all cases the use of reasoning gives the power to manipulate the environment to the animal's advantage (as with chimpanzees piling up boxes to reach bananas hung from the roof of their cage). In the case of man this manipulation of the environment has become very elaborate and this is the immediate reason for his remarkable success as an animal.

CO-ORDINATION IN PLANTS

38. Phytohormones. No plants above the level of algae or the gametes of bryophytes and pteridophytes have systems of co-ordination which can be compared with the nervous systems of animals.

Instead plants are co-ordinated by the activities of certain chemicals called *phytohormones*. In many ways these are similar in activity to animal hormones, being secreted in one part of the plant and then translocated to a target region of response. They are also active at very low concentrations.

A great deal of current botanical research is into the detailed interaction of the plant hormones and in studying the following it should be appreciated that our knowledge of these substances and their activities is still very incomplete.

39. Sub-divisions of phytohormones. Phytohormones can be sub-divided into the following:

(*a*) *Auxins.* These are to do with *growth* and particularly in the longitudinal growth of the plant body. Auxins are involved in the tropisms of plants, secondary growth and wound healing, the initiation of roots and inhibition of buds as well as with activities connected with fruit formation and abscission. Auxins work in conjunction with other types of hormones.

(*b*) *Gibberellins.* At first these seemed to be botanical curiosities but there is now a great deal of evidence that they are essential to the *overall co-ordination* of the plant. They work with auxins and flowering hormones.

(*c*) *Phytochromes or florigens.* These are hormones which control the *flowering times* of plants. Although quite a lot is known about their mode of operation they have yet to be chemically isolated (*see* **46**).

(*d*) *Kinins.* These stimulate mitosis and are important in differentiation.

PLANT TROPISMS AND THE ROLE OF AUXINS

40. Tropisms. These are the responses made by different parts of the plant body to unidirectional stimuli. The tropism may be positive or negative or at right angles, and the name of the stimulus which causes the response is used as a prefix. Thus a stem, growing towards the light is showing a positive phototropism. Primary stems are negatively geotropic. Roots are negatively phototropic and positively geotropic and possibly positively hydrotropic as well.

Secondary shoots and roots and leaves may show diatropisms and grow out at right angles from the plant body. In the case of the diaphototropic response of leaves hormones have been shown to be produced in the lamina and translocated to the region of response which is the petiole.

41. The source of auxins. The young cereal seedling first leaf (coleoptile) has been used extensively in auxin experiments. If the root or stem tip is excised (or covered in the case of light) no bending takes place when the plant is subjected to a unidirectional stimulus.

If the excised tip is placed on a small block of agar jelly the latter becomes active in the sense that it will cause bending if placed excentrically on the cut stem or root.

The conclusions from these and similar experiments are that the auxins are synthesised in the meristems just behind the root and stem tip. They are then translocated to the elongating regions some centimetres from the tip.

42. The chemical nature of the auxins. Auxins were first isolated from animal urine and at the present time a number of natural and artificial products exist.

The most important naturally occurring auxin is *indolyl-acetic acid* (or IAA) whose formula is

IAA is derived from the amino-acid tryptophan

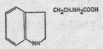

43. Formation of different concentrations of auxin in response to unidirectional stimuli. In order for the region of response to produce a bending movement it is necessary for different concentrations of auxin to be present on each side.

As far as the response to light is concerned, it seems that light actavites riboflavin enzymes (*see* VIII, **5**) which lead to the degradation of the IAA to an inactive product, i.e.:

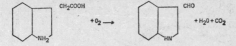

This happens most rapidly on the illuminated side so that differential concentrations of IAA are built up. Carotene may also be important in the receptor system possibly by absorbing light of the same wavelength as riboflavin in the tip and preventing the latter oxidising the auxins.

For responses to gravity it is likely that the stimulus of the gravity leads to an accumulation of auxin and/or auxin-producing bodies on the lower side of the stem or root tip.

The importance of lateral translocation of auxins in building up differential concentrations is not certain. It is probably a minor effect.

In both cases the auxin is found at twice the concentration on the one side of the plant as compared with the other.

44. How the auxin causes bending. From the above it can be seen that we have a situation in the plant where in the case of unidirectional light there is twice as much auxin on the dark side. In the stem auxin normally causes increase in cell size so the stem bends towards the light.

The situation in the root is that the auxin is concentrated in the lower part of the responding region but in this case the auxin inhibits cell elongation so the root bends downwards.

The opposite responses of roots and stems to the same concentrations of auxin can be seen in Fig. 127.

It is likely that the increase in cell size brought about by auxins is mainly due to changes in the rigidity of the cell wall. If this is decreased the wall pressure will fall so water will be taken into the cell by an increased water pressure deficit. The exact relationship between the properties of the cell wall, the auxin and the general metabolism of the cell is almost certainly very complex.

NOTE: The different responses of monocotyledons and dicotyledons to auxins is the basis of selective weed killers. The dicotyledons are much more sensitive and thus dicotyledon weeds amongst cereals and grass crops can be destroyed.

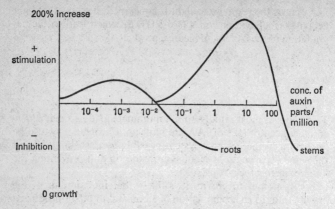

FIG. 127 *The responses of roots and stems to different concentrations of auxins.*

45. Summary of auxin and hormone activity. This is illustrated in Fig. 128.

46. Florigens and the control of flowering. The phenomenon whereby plants come into flower at certain times of year has been linked to the stimulus of light and is therefore termed photoperiodism. Flowering involves the plant hormones called florigens.

It is beneficial to a given species that all its members come into flower at more or less the same time of year in order that cross pollination should be successful. Many plants that originate from temperate regions such as the cereals e.g. barley are brought into flower by the accumulated effect of the light from the long days of summer. Such species are called long day plants (or LDP). Others originating from the tropics are stimulated to flower by short days and recurrent periods of darkness and consequently are called short day plants (or SDP). An example is the Chrysanthemum. Some other plants will flower under any conditions of light and are called neutral day plants (or NDP).

Experiments involving the exposure of mature leaves to different periods of daylight or the removal of mature leaves indicate clearly that determination of flowering comes from the leaves themselves. The system is thought to work by a series of phytochromes, coloured pigments in the leaves that exist in various

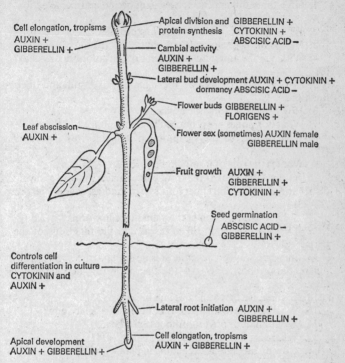

FIG. 128 *Summary of plant auxin and hormone activity.*

forms. Under conditions of long periods of light especially in the orange range the phytochrome P turns to PLDP while in dark or in blue light the phytochrome changes to PSDP. These light-sensitive phytochromes are in turn linked to the synthesis of the actual florigen hormones, also made in the leaf, but this time translocated to the flowering axes where they cause differentiation to occur.

Thus it is surmised that the plant has a biological clock which can measure the cumulative effects of light falling on it and can bring about flowering as may be appropriate. This is why flowers of a given species come out at the same time of year.

PROGRESS TEST 11

1. Give three differences between endocrine and nerve action. **(1)**

2. What structures are to be seen in a medullated nerve? **(4)**

3. How is the nerve impulse transmitted? **(5)**

4. What are the antagonistic systems of the autonomic system? **(8)**

5. List the main parts of the mammalian brain and their functions. **(9–12)**

6. How may receptors be classified? **(16)**

7. What hormones are produced by the pituitary? **(19)**

8. What would be the symptoms of an excess of thyroxin? **(19)**

9. What is the role of insulin in the body? **(19)**

10. Draw a diagram of the front end of an earthworm showing distribution of the central nervous system and associated nerves. **(22–25)**

11. What is the function of the giant fibres in worms? **(23)**

12. Describe the brain of an insect and the functions of the parts. **(27)**

13. In what way do juvenile hormones and the growth and differentiation hormones act together in insect development? **(31)**

14. Distinguish between taxis and kinesis. **(32)**

15. What are the main features of instinctive behaviour? **(35)**

16. What kinds of phytohormones exist? **(39)**

17. Where are auxins produced in plants? **(41)**

18. How do auxins cause bending in plants? **(44)**

EXAMINATION QUESTIONS

The figures in **bold** type indicate the marks allocated to each question or part question.

1. Give an account of **one named** animal hormone and **one named** plant hormone (in your answer you should consider where each is produced, how it moves within the organism, its mode of action and its role).

Do you consider that it is helpful to group both under the heading of *hormones* or do you think that, in the case of the plant, *growth substance* is a better term?

Cambridge, 1976

2. The apparatus shown below was used to demonstrate peristalsis in a portion of small intestine removed from a rat. It was also used to investigate the effects of specific chemicals on this process.

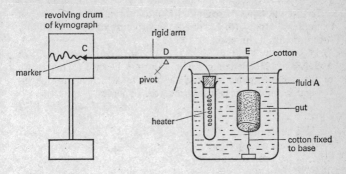

The traces obtained on the kymograph were as follows

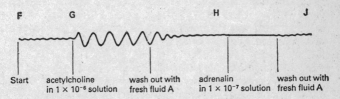

(a) What properties would the fluid **A** have, and why?

(b) What is the point of having a heater in the jar?

(c) Why is the length **CD** much greater than **DE**?

(d) Interpret the trace between **F** and **G**, explaining your reasoning.

(e) Interpret the trace between **G** and **H**, explaining your reasoning.

(f) Interpret the trace between **H** and **J**, explaining your reasoning.

(g) Suggest *two* ways in which the apparatus might be improved.

Associated Examining Board, 1975

3. What are the biological advantages of a sense of hearing in mammals? Make fully labelled diagrams to illustrate the structure of the ear of a mammal. Give a concise account of the way in

which sound vibrations are translated into nerve impulses.

London, 1976

4. Give an account of the ways in which nerve impulses are conducted by the nervous system.

London, 1975

5. List the types of receptors found in organisms and give an example of each. **(6)** Correlate the structure of any *one* animal receptor with its function. **(14)**

Oxford and Cambridge, 1976

6. Make a labelled diagram of a neurone. Discuss (*a*) the nature of the nerve impulse and the methods by which it may be investigated and (*b*) the general properties of nerves.

Oxford and Cambridge, 1973

CHAPTER XII

Effector Systems in Locomotion

1. Introduction. The response to stimuli that we have been considering in the previous chapter may take the form of secretory activity by glands or by operation of the effector systems involved in locomotion.

In this chapter certain aspects of the working of three types of effector systems will be considered. These are the *striated muscles of vertebrates*, *cilia* (and *flagella*) and *pseudopodia*. While the latter types of locomotion characterise groups of the protozoa all three types are found in many higher animals (*see* Table XXII).

TABLE XXII.   TYPES OF MOVEMENT FOUND IN MAN

Types of movement	Examples
Generalised movements	muscles operating across joints.
Movements of solids of small size	ciliary action in trachea and Fallopian tubes.
Movements of whole cells	flagella of spermatozoa. Pseudopodia of certain white corpuscles.

MUSCLES

2. The anatomy of striated muscle. There are well over one hundred different muscles in the human body and these are mostly attached from one skeletal element to another, across joints. Tendons are the specialised connective tissues which fix a muscle to its bone.

(*a*) *The fibre.* Individual muscles are made up of many fibres which vary from 1–40 mm in length and from 10–40μm in width. The fibres are each made of many cells and are syncytial, with a number of large nuclei at the periphery. Around the outside of the fibre is the sarcolemma (muscle membrane) and to this is fixed the motor end plate from a nerve. The sarcoplasm of the fibre contains a large number of mitochondria and an elaborate endoplasmic reticulum but is mainly composed of the myofibrils whose protein systems form the basis of muscular contraction.

(b) *Patterns of the striations.* The pattern of protein distribution in the myofibrils is reflected in the whole striations of the fibre. There are sharply defined Z lines in the centres of the clearer I bands and these alternate with denser A bands. A clearer area within the latter is the H band (*see* Fig. 129).

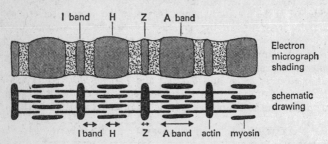

FIG. 129   *Mammalian striated muscle showing the pattern of banding and of the myosin and actin distributions.*

(c) *Muscle proteins.* The two proteins that are concerned in the contraction mechanism are actin and myosin. The distribution of these proteins is shown in Fig. 130.

The two kinds of filaments are linked together by a series of cross bridges which project outwards at regular intervals of 6–7 nm and with each bridge separated by 60° from its neighbour. The bridges thus form a helical pattern which repeats itself every six bridges.

3. The mechanism of contraction.

(a) *The sliding filaments.* Observations of the changes in band pattern during contraction and relaxation of striated muscle have led to various theories as to the mechanism. The most generally accepted theory is that of H. E. Huxley who suggested that the myosin bridges are able to oscillate backwards and forwards and link up with specific sites on the actin filaments. The bridges would then contract and thus pull the actin into the myosin, immediately becoming released from the latter and reaching out again for a new link further along the filament.

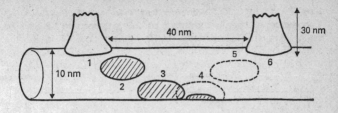

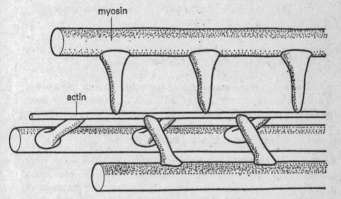

FIG. 130 *Cross bridges between myosin and actin in muscle.*

This theory explains the observation that during muscle contraction the I bands shorten. Experimental observations on the utilisation of ATP by contracting muscles on the basis of one ATP being split for every bridge made, showed that a single bridge might go through between 500 and 1,000 cycles of extension, junction and contraction in one second.

(b) *Basis of contraction.* The nerve impulse spreading from the nerve motor end plate spreads across the muscle and causes changes in the permeability of the membranes in which Ca^{++} is bound in vesicles. The Ca^{++} diffuses into the regions of the filaments and there causes (via a complex sequence of changes involving ATP) the reversal of the charge at the end of the cross bridge. These now extend pushing the actin rods further into the myosin and thus bringing about overall contraction of the muscle (*see* Fig. 131). (This is very controversial, many authorities believe a contraction of cross bridges is the active part of the movement.)

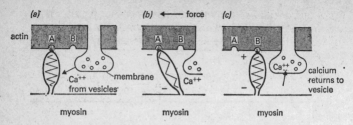

FIG. 131   *Action of cross bridges in muscle contraction.*

(*a*) Muscle at rest: opposite charges at ends of cross bridges. (*b*) Muscle contraction: at the ends of the cross bridges the charges are now the same, causing the bridge to expand. (*c*) Muscle recovery: cross bridges contract and make new linkages at B.

ATP is again required to release the actin–myosin linkage and to provide energy to pump the Ca^{++} back into the vesicles ready for the next contractions. Where calcium metabolism goes wrong in an animal, due for example to loss or malfunction of the parathyroids, the muscles lock into a state of tension.

4. The nerve impulse and muscular contraction. The nerve fibres contract only when they receive impulses from the motor end plates of the neurones. The relationships of the end plates to the fibre are shown in Fig. 109 and each single neurone may have well over one hundred of these end plates to different muscle fibres.

It is thought that the arrival of an impulse at the surface of the muscle sets off an action potential along the muscle itself, this potential, like that of neurone, depending on Na^+ and K^+ exchanges across the muscle membrane.

5. Tetanic contraction. As with the neurone the muscle fibre shows the phenomenon of absolute and relative refractory periods. During the former an impulse arriving at the muscle will cause no contraction while during the latter the contraction will be reduced. If a large number of impulses arrive close together the muscle shows a tetanic contraction.

The actual power of contraction of a given muscle depends on the number of motor units that are being stimulated. Any one muscle, say the biceps, has a very large number of these units each with its separate innervation. If a small load is placed on the muscle only a few units are contracted whereas a large load leads to the operation of all the units. At any one time proprioceptors within the muscle feed back to the spinal cord information about the weight being applied and thus to the numbers of units required.

6. The energy supply for muscle contraction. In a simple form the energy supply for muscle contraction may be shown as in Fig. 132. While this scheme indicates the vital stages of the energy exchanges it is very much oversimplified.

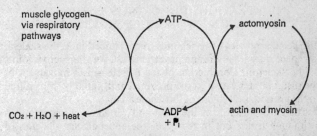

FIG. 132   *Simplified diagram of energy supply for muscle contraction.*

A more exact picture is shown in Fig. 73. The down grading of the respiratory substrate such as glycogen leads to the formation of ATP and this is maintained in the muscle at a very low level, enough for only eight contractions. The ATP exists in a relationship with another energy-store molecule called creatine phosphate (CP):

$$ATP + C \rightleftharpoons ADP + CP.$$

As soon as the ATP concentration rises in the muscle this reaction goes to the right and a large quantity of CP is formed, enough for one hundred contractions. When the muscle actually contracts it draws on the ATP supply which is immediately replenished from the CP store. In other words the readily available energy supply of the muscle is not ATP but CP.

If the CP store is diminished it has to be replenished from the respiratory breakdown of glycogen (*see* VIII, **5**). To do this aerobically takes some time and the muscle may obtain its energy from the first anaerobic stages of respiration. These result in accumulations of *lactic acid* being formed and in the body building up an oxygen debt. This lactic acid tends to be transported to other parts of the body such as the liver and heart muscles for resynthesis or breakdown. It cannot be handled by the muscles themselves.

A very fit human being can generate 4g of lactic acid in the whole body in a second and may accumulate as much as 120g before he is brought to a standstill. This 120g represents a very large oxygen debt of some twenty litres and it may take many minutes of strained breathing to repay it.

THE RELATIONSHIPS OF MUSCLES TO THE SKELETON TO PRODUCE MOVEMENT

7. Skeletal muscles of vertebrates. As explained in **2**, the skeletal (striated) muscles are found in vertebrates operating across joints of the skeleton. Contraction of the muscle causes movement of the distal bone relative to the proximal and, suitably transmitted to the ground, movement of the whole animal.

The axial skeleton of the head and vertebrae is important in swimming forms but in tetrapods the appendicular skeleton of the limbs has taken over the role of locomotion.

The appendicular skeleton consists of the limb girdles and humerus, radius and ulna, and carpals of the forelimb and femur, tibia and fibula and tarsals of the hindlimb.

8. Movements about the shoulder and thigh joints. The movements of the proximal part of the limb is produced by sets of muscles between the limb girdle and the first limb bone. The muscles operate in antagonistic pairs:

- (a) Pair 1: Forward movement—protractor;
  Backward movement—retractor.
- (b) Pair 2: Upward movement—elevator;
  Downward movement—depressor.
- (c) Pair 3: Rotary muscles moving bone clockwise and anticlockwise.

For walking the first limb bone is drawn backwards in the power stroke and forwards in the recovery movement while for flying the limb is depressed and rotated in the power stroke and elevated and rotated in the recovery stroke.

9. Movements within the limb. The six types of movement seen in the first part of the limb are combined with other movements within the limb itself to provide efficient locomotion. The joints of the lower parts of the limb are the elbow and wrist and knee and ankle and these tend to be hinge joints allowing movement in one plane.

The muscles across these joints act in antagonistic pairs:

(a) *Hindlimb:*
  (i) Across the knee: extensor of the tibia;
                        flexor of the tibia.
  (ii) Across the ankle: extensor of the foot (tarsals);
                         flexor of the foot.
(b) *Forelimb:*
  (i) Across the elbow: extensor of the radius and ulna;
                        flexor of the radius and ulna.
  (ii) Across the wrist: extensor of the hand (carpals);
                         flexor of the hand.

In walking (of a four-footed animal), the power comes from the retraction and extension of the tibia, radius and ulna, and tarsals and carpals. The recovery stroke is the flexing of these bones.

At the end of the tarsals and carpals there will be some means of ensuring a firm grip on the ground as the whole limb retracts and extends (hooves, claws, etc.). Increase in speed is brought about by a lengthening of the bones of the limb and especially the digits so that a wider arc is traversed at each pace. The powerful muscles of the backbone also assist in straightening the vertebral column and thus increasing the thrust developed against the ground.

10. Skeletal muscles in arthropods. Unlike the endoskeleton of the vertebrates, that of arthropods is an exoskeleton and the muscles are contained inside. The principle of their operation is substantially the same as the vertebrates and antagonistic pairs of muscles are found running within the joints and are attached to the chitinous projections or apodemes of the exoskeleton.

The joints of the exoskeleton are made of *flexible chitin* and are often protected by hard spines from the tough exocuticle. Having a muscular system contained within the dead exoskeleton has the advantage of giving protection to the animal. Its disadvantages are that it restricts the size of the animal (as the strength of the muscles and the weight of chitin that can contain them do not increase proportionally) and that extensive *moulting* must occur before growth.

(*a*) *Insect flight.* Here small movements of the insect thorax lid (the tergum) against the base (the sternum) are magnified to produce very extensive movement of the wing tips. These thoracic movements are due to indirect flight muscles, namely the circular and longitudinal muscles of the thorax. Direct flight muscles attached to the bases of the wings ensure the rotation of these during flight to provide the maximum (downstroke) or minimum (upstroke) surface to the air.

(*b*) *Insect walking and running.* Insects use six legs for walking and running. These operate as two stable tripods such that the first and third leg on the left side are pushing against the ground at the same time as the second leg on the right side. Meanwhile first and third legs on the right and second leg on the left are all moving forwards at the same time. In this way the insect rotates slightly from one tripod to the other as it moves.

Like some tetrapods the faster insects have extended legs with the main muscle power concentrated in the first section or coxa. The five-jointed tarsus of the feet ends in claws.

OTHER EFFECTOR SYSTEMS

11. Cilia and flagella. Cilia and flagella are organelle structures which project from certain cells and are concerned with *locomotion in a liquid medium*. Whereas cilia are usually numerous and short (5–10μm) flagella are *few and long* (up to 150μm). The structures have the same basic parts and operate in the same way.

Both cilia and flagella are widespread in living organisms. Some instances are as follows:

(*a*) *Animals:*

(*i*) Protozoa: flagella for locomotion; cilia for locomotion and feeding currents.

(*ii*) Coelenterata: e.g. *Hydra*, digestive activity partly due to cilia.

(*iii*) Platyhelminthes: e.g. *Planaria*, cilia involved in locomotion.

(*iv*) Annelids: e.g. *Lumbricus*, cilia produce currents in nephridium; larval polychaetes swim by ciliary action, cilia also provide their feeding mechanisms.

(*v*) Chordates: spermatozoa use flagella for locomotion, cilia are commonly found in oviducts, in the trachea, etc. There is a ciliated epithelium.

(b) *Plants:*

(*i*) Algae: flagella used in locomotion of unicellular algae such as *Chlamydomonas, Euglena* and *Volvox*. Flagella propel antherozoids of higher algae such as *Fucus*.

(*ii*) Fungi: zoospores, as in *Phytophthora*, swim with flagella.

(*iii*) Bacteria: many bacilli and spirochaetes have flagella.

(*iv*) Higher plants: antherozoids in Bryophyta, Pteridophyta and certain primitive Spermatophyta such as *Gingko* have flagella for locomotion.

12. Relationship of size of organism to medium. Where an organism is very small such as a spermatozoan, antherozoid, zoospore or unicellular animal or plant the activity of cilia, or flagella, is sufficient to cause it to move through the liquid medium in which it lives. These organelles are effective means of propulsion and, as in the case of ciliates such as *Paramecium*, may allow their owners to travel at considerable speed (in terms of traversing their own lengths per minute).

If cilia were anchored down on an epithelium they would cause the movement of a liquid medium or particles of solid in that liquid. This fact is exemplified in the movement of eggs down an oviduct, or of particles up a trachea or of a current of fluid through nephridium, or in filter-feeding mechanisms of certain bivalves, polychaetes and other animals.

13. Ultrastructure of cilia and flagella. While some observations, from the use of light microscopes, have been made on the tendency of sperm tails to split, the structure of cilia and flagella was not understood until the advent of the electron microscope.

It was found that an outer membrane covered the structure and that this was an extension of the cell membrane. Within this sheath was a collection of nine peripheral fibrils and two central fibrils. The outer fibrils were double and the central single (*see* Fig. 133). There is a basal granule (kinetosome) from which the cilium or flagellum develops. The centriole of the nucleus is homologous to a kinetosome.

14. Mechanism of action. The power stroke of the cilium is to bend stiffly and fast in one direction and then to recover by a slower process involving some bending in the middle (*see* Fig. 133). The power stroke exerts more pressure on the environment than the recovery stroke because of the decrease in the surface presented.

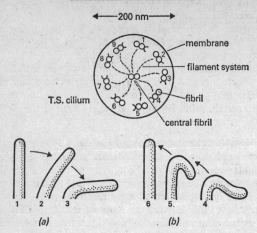

FIG. 133   *The ultrastructure and method of contraction of a cilium.*

(*a*) The power stroke: fibrils on the shaded side contract rapidly. (*b*) Recovery stroke: fibrils on the shaded side relax; fibrils on the non-shaded side contract slowly, starting at the base.

It has been suggested that the impulse for contraction is passed rapidly up the central fibrils and out to fibrils numbers 1, 9, 2, 8 and 3. Connections between the central and peripheral fibrils are visible. These five fibrils then contract more or less simultaneously and the cilium performs its power stroke.

Meanwhile a slower set of impulses is passing up the fibrils numbers 7, 4, 6 and 5 which start to contract progressively along the length of the cilium starting at its base. The effect of this is to produce the recovery stroke.

Flagella act in the same sort of way but can produce waves of contraction rather similar to the swimming action of an eel. The movement of these waves backwards against the pressure of the environment causes the organism to be driven forwards.

15. Co-ordination of cilia. Cilia normally beat in a *metachronal rhythm* with waves passing along the ciliated surface (described as similar to a gust of wind over a corn field). For this rhythm to operate cytoplasmic continuity is necessary.

It is suggested that the impulse initiated by one basal granule passes up its own cilium but also triggers off an impulse in the next basal granule some milliseconds later.

16. Amoeboid movement.

(*a*) *Occurrence.* Amoeboid movement is found in the rhizopod protozoa such as the amoeba species, in the slime fungi, in the white corpuscles of higher animals and in the spermatozoa of nematodes. It may also be similar to the mechanism of protoplasmic streaming observed in many plant cells.

Large amoeba may move up to $4\mu m$ per second by this method of locomotion and white corpuscles at some $0\cdot5\mu m$ per second. The method is suitable for moving over a substratum or along a duct. It is interesting that the soil "amoeba" *Naegleria gruberi* can be induced to retract its pseudopodia and put out a flagellum under suitable conditions.

(*b*) *Possible mechanisms of amoeboid movement.* The main and obvious feature of movement within a pseudopodium or whole amoeba is that there is a central core of liquid plasmasol (or endoplasm) moving in the direction of locomotion. At the same time there is an outer tube of more viscous plasmagel (or ectoplasm) which does not appear to move.

There are three possible explanations (*see* Fig. 134).

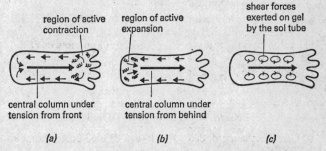

FIG. 134   *Three possible mechanisms of amoeboid motion.*

(*i*) The sol is changing to gel at the front of the pseudopodium and at the same time contracting. There is a slow counterflow of gel in the reverse direction. The animal is thus drawn forwards from the front.

(*ii*) The gel is changing to sol at the hind end of the organism by an active process so that a stream of sol is being pushed forwards from the rear.

(*iii*) The central sol is being driven along the gel tube by shear forces acting along the length of the tube.

(c) *Summary of theories*. There is experimental evidence to support each of these explanations of amoeboid movement. It is also clear that ATP is involved in the process as well as proteins equivalent to the muscle proteins of higher animals.

At present it is not possible to say which of the explanations is correct or whether there are several different types of amoeboid and streaming movements in different organisms.

PROGRESS TEST 12

1. What is an effector organ? **(1)**
2. Which structures are found in a muscle fibre? **(2)**
3. How are the proteins actin and myosin distributed in a fibre? **(2)**
4. What role does ATP play in contraction? **(3)**
5. What change does the nerve impulse cause in the muscle? **(4)**
6. What is creatine phosphate? **(6)**
7. Which pairs of muscle may operate across a joint? **(7)**
8. What adaptations of the limb bones and muscles are associated with running in tetrapods? **(9)**
9. What are the limitations imposed by an exoskeleton? **(10)**
10. Where are cilia and flagella found? **(11)**
11. What does the transverse section of a cilium show? **(13)**
12. Draw the stages in the beat and recovery of a cilium. **(14)**
13. Which are the main features of amoeboid movement? **(16)**

EXAMINATION QUESTIONS

1. Describe the methods by which (a) a ciliated protozoan and (b) an amoeba move from place to place.

How is locomotion of survival value to each of these organisms?

after London, 1976

2. Discuss the role of ATP in muscle contraction.

after Oxford and Cambridge, 1976

PART THREE

GENETICS, EVOLUTION AND ECOLOGY

CHAPTER XIII

Mendelian Genetics and the Exceptions to Mendel's Laws

1. Definitions. Genetics is the *study of heredity*. The first successful quantitative studies were made by *Gregor Mendel* (1866). Mendel was able to show that the characters of an organism are determined by *specific germinal units*, or *genes* as we call them today, and that these genes are handed down in an exact manner from parent to offspring. He also showed that *genes do not blend* but remain quite discrete from one generation to another.

In the present chapter Mendel's two famous laws of *quantitative inheritance* will be considered and then the various exceptions that exist. The important modern synthesis between genetics and evolution is discussed at the end of the following chapter (*see* XIV, 19–31).

The terms commonly used in genetics are as follows:

(*a*) *Gene:* a part of a chromosome which either itself or together with other genes is responsible for determining a character of the individual possessing the gene.

(*b*) *Dominant genes:* those which manifest their character when present in the heterozygote condition (*see* 2(*c*)).

(*c*) *Recessive genes:* those which only manifest their character when present in the homozygote condition.

(*d*) *Heterozygote* (adj.-ous): an individual who has inherited dissimilar genes from each parent.

(*e*) *Homozygote* (adj.-ous): an individual who has inherited similar genes from each parent.

(*f*) *Locus:* a site on a chromosome occupied by a gene.

(*g*) *Alleleomorphs* (*or alleles*)*:* genes having contrasting effects situated at the same locus.

(*h*) *Genotype:* the genetic constitution of an individual.

(*i*) *Phenotype:* the characters actually displayed by the individual (i.e. recessive genes present will not show up).

(*j*) *Linked genes:* those situated on the same chromosome.

(*k*) *Sex-linked genes:* those which are situated on one of the X or Y sex chromosomes.

(*l*) *Filial* 1 (F.1): the first generation of a cross.

(*m*) *Filial* 2 (F.2): the second generation of a cross.

(*n*) *A back cross:* a cross between the heterozygote offspring and the homozygote parent carrying recessive genes.

2. Mendel's first law of definite proportions (monohybrid inheritance).

(*a*) *The law.* Mendel's first law states that when two purebred individuals showing a pair of contrasting characters are crossed then the characters will *segregate out in definite proportions* in the second filial generation.

(*b*) *How the law was established.* Mendel worked largely with the pea plant (*Pisum*) as this plant is normally self-pollinated, so that he could be sure it would breed true. When he wished to make a cross between plants, the stamens could be removed from the flower to be pollinated and foreign pollen deliberately brushed on to the receptive stigma.

The characters used by Mendel are shown in Table XXIII.

TABLE XXIII.   CHARACTERISTICS OF *Pisum* USED TO SHOW MONO-
HYBRID INHERITANCE

Characteristics	Dominant	Recessive
Size of plant	tall	short
Colour of cotyledons	yellow	green
Seed coat	round	wrinkled

Mendel made the crosses between plants that had bred true for a number of generations (*see* Fig. 135). It should be noted that these ratios were obtained using hundreds of individuals. As more are used so the ratios in the F.2 approach the expected frequency of 3:1.

(*c*) *The use of symbols to represent genes.* In the crosses shown in Fig. 135 the parents are homozygous and are, of course, diploid. The dominant genes are those giving tallness, yellowness and

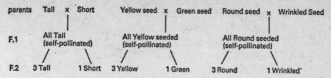

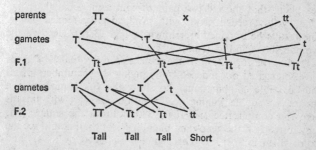

FIG. 135 *Three of the contrasting character pairs used by Mendel and the results of F.1 crosses of pure-bred individuals.*

round seeds respectively (this is by definition since the F.1 progeny must clearly be heterozygous). It is customary to give the capital initial letter of the character to the dominant gene and the small letter to the recessive.

With this information it is therefore possible to represent the three crosses in symbolic form, and also to indicate the gametes which are haploid, and their various combinations. Figure 136,

FIG. 136 *The F.2 generation produced by crossing tall and short pure-bred pea plants indicating the segregation of genes determining the characters.*

shows the cross between the tall and short plants, while Fig. 137 indicates the cross between the yellow and green-seeded plants. The student should be able to work out the symbols for the third cross between the round and wrinkled-seeded plants for himself.

3. Mendel's second law of independent segregation.

(a) *The Law.* Mendel's second law states that when two pure-bred individuals showing two or more pairs of contrasting characters are crossed then the characters *segregate out independently* in the second filial generation

(b) *How the law was established.* Mendel crossed pure-bred

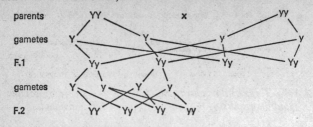

FIG. 137 *The F.2 generation produced by crossing yellow and green cotyledon pea plants indicating the segregation of genes determining the characters.*

pea plants that showed two pairs of contrasting characters, i.e. round yellow-seeded plants with wrinkled green-seeded plants. The results he obtained are shown in Fig. 138(*a*). He used the symbols shown in Fig. 138(*b*).

During the formation of the gametes the two pairs of genes have segregated out independently giving equal numbers of all four possible combinations. To show the F.2 composition a square is made, as joining up the various gametes with lines is here too complicated (*see* Fig. 138(*c*).) From the square it is easy to read off the geno- and phenotypes of the F.2 and see the ratio which is 9:3:3:1. This figure is obtained by $(3:1)^2$ and indicates independent segregation of the genes.

If individuals showing three pairs of contrasting character were crossed the ratio of the F.2 progeny would be $(3:1)^3$ which is 27:9:9:9:3:3:3:1.

4. The back cross. An application of Mendel's laws is found in the *back cross* of an individual of unknown genotype with the double recessive individual to establish the genotype of the former.

Thus if we have a tall pea plant of unknown genotype (i.e. it could be Tt or TT) and we cross it with a short plant whose genotype must be tt, the ratio of the offspring will depend on the original unknown parent (*see* Fig. 139(*a*) and (*b*)).

5. Exceptions to Mendel's laws. Although Mendel's laws are fundamental to the understanding of genetics there are a number of exceptions to them which have been discovered since 1900.

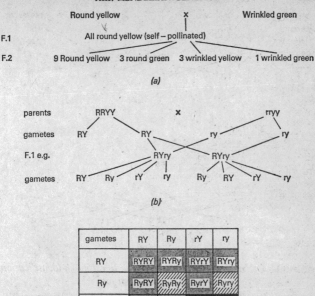

FIG. 138 *The independent segregation in the F.2 generation produced by crossing individuals with two pairs of contrasting characters.*

(*a*) *Partial dominance.* Here we have two alleles of which neither is dominant so that the F.1 offspring shows intermediate characters between the parent. Well-known examples are the pink flowers produced by breeding red and white snapdragons and the speckled chicks from the cross of black and white Andalusian fowls. For such genes we must use two different symbols as capital and small letters would be inappropriate.

The crossing of two heterozygotes for partial dominance, say in the snapdragon, would work out as shown in Fig. 140.

(*b*) *Lethal genes.* Certain combinations of (normally) recessive genes could produce lethal characters where the offspring

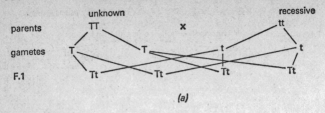

(a)

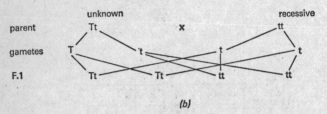

(b)

FIG. 139 *Use of the backcross technique to establish identification of an unknown genotype.*

(*a*) All the F1 generation were tall, implying that the unknown was TT.
(*b*) The F1 generation had a ratio of 1:1 tall:short, implying that the unknown was Tt.

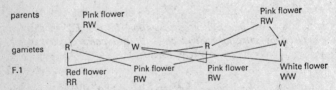

FIG. 140 *An example of partial dominance in the heterozygote.*

never develops or begins to develop but aborts. Such is the case when two haemophiliac genes come together in the human female and lead to the death of the early foetus.

(*c*) *Linkage.* As most species have many chromosomes and the genes are found distributed along the length of the chromosomes, genes that are on the same chromosome are linked together and thus may not segregate independently in meiosis.

In the gamete formation instead of four combinations of the

two genes concerned, only two types are produced. The banana fly, *Drosophila melanogaster*, has only four chromosomes and the genetics of this insect have been very extensively investigated. The normal body (G), grey, exists as a recessive form called ebony body (g) and the normal antenna (A) exists as a mutant form (a) which is similar to a small extra leg. This form of antenna is called aristopedia. These two different pairs of genes are both on the same chromsome and Fig. 141 illustrates the re-

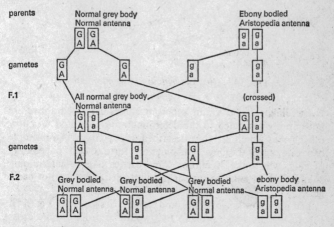

FIG. 141 *Segregation of linked genes does not take place, as in Mendel's second law.*

sults of crossing pure-bred flies showing such linked characters. This gives the monohybrid ratio of 3:1 and not the dihybrid ratio of Mendel, i.e. 9:3:3:1. The boxes indicate linkage.

As we shall see below (8) cross over of linked genes can produce recombinations of character and this phenomenon is used to map the position of the genes along the chromosome.

SEX CHROMOSOMES

6. Inheritance of sex. The determination of an individual's sex is initially due to the inheritance of whole chromosomes called X and Y. In some animals the male is the heterogametic sex having the genes XY while the female is the homogametic sex having

XX. This is the case with mammals. In birds the position is reversed and the males carry XX while the females are XY.

It is clear that this system tends to produce equal numbers of male and female offspring. Thus in the case of man there is the result shown in Fig. 141.

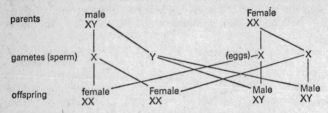

FIG. 142   *Sex determination in mammals.*

7. Sex linkage. The X and Y chromosomes described above are not entirely restricted to the determination of sex and do carry other genes which are thus described as *sex-linked*. An example of sex linkage is the ginger and black and tortoiseshell coat colours of cats, where the X chromosome carries a gene-determining colour. The Y chromosome is usually inert and does not carry genes.

The combination of sexes and colour is shown in Fig. 143. Figure 143 shows the progeny that would be produced if a cross

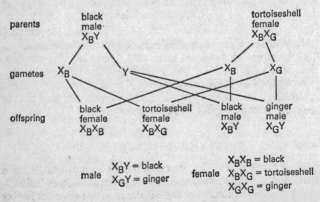

FIG. 143   *Behaviour of sex-linked genes and the coat colour of cats.*

were made between a black male and a tortoiseshell female. There are many examples of sex linkage in man including the harmful genes that cause haemophilia and colour blindness.

GENETIC RECOMBINATIONS

8. Cross-over frequency and chromosome maps. As seen in **5** linked genes are found on the same chromosomes and tend not to segregate at meiosis. However during the prophase of meiosis the crossing over of genetic material between pairs of homologous chromosomes may take place.

Whether or not two linked genes become recombined in this way depends largely on their distance apart along the chromosome. Genes that are very close seldom cross over whereas those at opposite ends of the chromosome may show a high cross-over frequency.

Cross-overs produce variations from the 3:1 ratios expected in the F.2 offspring of individuals with pairs of linked genes. In the case of *Drosophila*, discussed above, cross-overs were ignored whereas in fact the two pairs of genes controlling body colour and development of the antennae cross over in some twelve per cent of the F.2 offspring. This means that the numbers of each type for 100 F.2 progeny might be 66:6:6:22 (*see* Fig. 144). With a cross-over value of twelve per cent the genes are described as being twelve Morgans apart on the chromosome. A "Morgan" is used as the unit in honour of the distinguished American geneticist of that name who performed the classical investigations on *Drosophila*.

By this method the relative positions of hundreds of genes of *Drosophila* and some other organisms have been determined.

9. The relation of genetics to the mechanism of evolution. Understanding the laws of heredity and the nature of variation and how it arises and is transmitted is an essential part of our knowledge of the mechanism of evolution.

The modern synthesis of Darwin's theory of natural selection and the knowledge of genetics is incorporated into *neo-Darwinism* which is the suggested mechanism of the evolutionary process. This topic is discussed at the end of the following chapter.

10. Some human genes and their inheritance. It is clear that humans are not suitable material for genetic experimentation. The crosses that a geneticist would wish to make do not necessarily

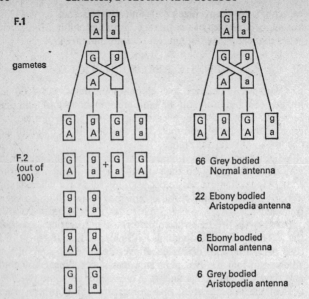

FIG. 144 *Percentage cross over of linked genes may be used as a means of determining their distance apart on a chromosome.*

occur, statistically-significant numbers are not available and the generation time is some twenty-five years. Despite these difficulties, genealogical trees have given quite a lot of information about the inheritance of some human characters, especially those causing hereditary diseases.

  (*a*) *Simple recessive genes:*
    (*i*) Albinism (no pigmentation).
    (*ii*) Amaurotic idiocy.
    (*iii*) Blue eyes.
    (*iv*) Hare lip.
    (*v*) Phenyl-thio-urea, a non-taster.
    (*vi*) Bifida of the spine.
  (*b*) *Simple dominants:*
    (*i*) The normal alleles of the above.
    (*ii*) Certain allergies.
    (*iii*) Diabetes (some forms only).

(*iv*) Night blindness.

(*v*) Extra teeth.

(*c*) *Sex-linked recessives:*

(*i*) Red-green colour blindness.

(*ii*) Haemophilia.

(*d*) *Sex-linked dominants:*

(*i*) Defective enamel of teeth.

(*ii*) Inborn nystagmus (giddiness).

The above are examples of unifactorial characters where one gene determines one character. Such things as viability, intelligence, fertility, height, weight, etc. are polygenically determined and produce continuous variation throughout the population.

PROGRESS TEST 13

1. What is meant by a "heterozygote", a "recessive gene", an "allele" and a "F.2 generation"? (**1**)

2. What is Mendel's first law? (**2**)

3. Why did Mendel choose the pea plant? (**2**)

4. Which of the following have dominant genes: tall plants, wrinkled seeds, green seeds? (**2**)

5. What is Mendel's second law? (**3**)

6. What is the use of a back cross? (**4**)

7. Give an example of partial dominance. (**5**)

8. Why are some genes linked and others not? (**5**)

9. How is sex determined? (**6**)

10. What progeny could result from crossing a ginger male cat and a tortoiseshell female? (**7**)

11. To what phenomenon is the recombination of linked genes due? (**8**)

12. If two genes show a high rate of recombination what does this indicate? (**8**)

13. Why does the inheritance of human intelligence not follow a Mendelian pattern? (**10**)

EXAMINATION QUESTIONS

The figures in **bold** type indicate the marks allocated to each question or part question.

1. A male specimen of *Drosophila* with fused veins and bobbed bristles was crossed with a female with normal veins and long bristles. Normal veins and long bristles are the wild type characters carried by dominant genes.

The progeny were of the following phenotypes:
   normal veins and long bristles,
   normal veins and bobbed bristles,
   fused veins and long bristles,
   fused veins and bobbed bristles.

(a) Suggest suitable symbols to represent the genes involved.

(b) What is the genotype of the female parent?

(c) What would be the expected ratio of these four phenotypes if the alleles involved showed independent assortment? Explain how this ratio would have been obtained.

(d) In practice, the number of progeny obtained were:

normal veins and long bristles	229
normal veins and bobbed bristles	18
fused veins and long bristles	17
fused veins and bobbed bristles	236

This indicates that the vein and bristle characteristics are determined by genes linked on the same chromosomes and that they were separated to a certain extent by meiosis during crossing over, thus giving rise to individuals showing recombination.

Draw a diagram to illustrate the parental chromosomes which gave rise to these progeny, showing them as they would appear during the latter part of the first prophase of meiosis.

(e) The number of progeny showing recombination, expressed as a percentage of the total offspring, is termed the crossover value for the two crosses involved.

Work out crossover values for the vein and bristle characters.

(f) Crossover values are used in the construction of chromosome maps: the closer the genes the less frequently crossing-over occurs between them. Using the crossover values you obtained in (e) and the crossover values stated below, draw a map of the chromosomes bearing the genes for bar eye, fused veins, and bobbed bristles.

bar eye/bobbed bristles	9%
bar eye/fused veins	2%

Associated Examining Board, 1977

2. Distinguish between the *genotype* and the *phenotype* of an organism. (5)

White face and long horns in a certain variety of cattle are inherited as single, dominant genes. A white-faced, long-horned bull was mated with a white-faced, long-horned cow. The calf had a white face and developed long horns. By means of diagrams, give the possible genotypes (in respect of face colour and horn length) of the bull, the cow and the calf. Describe the tests you would make to distinguish between the alternatives you suggest. (13)

Cambridge, 1976

3. Plants of *Mirabilis*, breeding true for both red flowers and broad leaves, were crossed with plants breeding true for both white flowers and narrow leaves. The result F_1 plants had pink coloured flowers and an intermediate leaf width. Suggest an explanation for this result.

What would be the expected appearance of the plants produced (a) by crossing the F_1 plants amongst themselves, and (b) by crossing the F_1 plants with the red-flowered, broad-leaved parental type? Give reasons for your answer.

Indicate how the results might have been different if the genes for flower colour and leaf width had been on the same chromosome.

London, 1975

4. Explain the meaning and discuss the genetical significance of the following: (a) segregation; (b) crossing over; (c) mutation; (d) dominance.

London, 1975

5. In tomatoes, tall plants are produced by the dominant gene D and its recessive allele d produces short plants. Hairy stems are produced by the dominant gene H and hairless (smooth) stems by its recessive allele h. Red stems are produced by the dominant gene A and non-red (green) stems by its recessive allele a.

(a) What are the possible genotypes of plants which are: (i) Tall, hairy and green-stemmed; (ii) Short, smooth and red-stemmed? (6)

(b) On crossing a plant which was short, hairy and green-stemmed, with one that was short, smooth and red-stemmed, the following progeny were obtained: 29 short, hairy, red; 25 short, hairy, green; 28 short, smooth, red; 30 short, smooth, green.

What were the genotypes of the parents? Give a reasoned explanation of your answer. (12)

(c) What is meant by a back cross (test cross)? Explain how such a back cross could be used to show whether a tall, smooth, green-stemmed plant was heterozygous or homozygous for the "tall" gene. (7)

London, 1976

6. What do you understand by linkage and sex linkage? Give an example of each and discuss the reasons why linkage is rarely complete.

Oxford and Cambridge, 1975

7. Discuss Mendel's laws of heredity. **(4)** Using examples show why these laws do not always seem to apply. **(16)**

Oxford and Cambridge, 1977

CHAPTER XIV

Evolution

INTRODUCTION

1. The concept of evolution. The theory of organic evolution implies that living organisms have arisen from earlier forms over very long periods of time. While some contemporary animals and plants resemble their ancestral types, most have substantially diverged from them.

(*a*) *Divergence and speciation as a natural process.* The majority of biologists believe that this process of divergence and speciation has taken place by the operation of natural processes which are still active at the present time. It is also considered that life itself arose by a process of gradual evolution from non-living matter.

(*b*) *Special creation.* Contrasting with the idea of evolution is that of *special creation* which supposes species to have been made originally by God according to the account in Genesis. While very few believe in the literal truth of the Biblical story of creation today it is worth remembering that it was a widely-held idea for a very long time and was only overthrown by the theories of Darwin and Wallace in the mid-nineteenth century.

(*c*) *Mechanism of speciation.* The names of *Darwin* and *Wallace* are associated with the theory of evolution because they suggested a mechanism whereby speciation could take place by natural processes. It is still convenient to consider evolution from two aspects, that is, to refer to the evidence by which the process has actually occurred and secondly to refer to the mechanism by which changes have been grought about.

The evidence for evolution embodies a classical range of geological and biological evidence which, while largely circumstantial, is certainly very convincing. The mechanism of evolution is less clear, however, and our knowledge at the present time is far from complete.

GEOLOGICAL EVIDENCE

2. Palaeontology. Palaeontology is the study of fossils and it provides a great deal of information not only that evolution has taken place but also about the major features of the succession of animal and plant life over the last five hundred million years (*see* 3–7 below).

3. Fossils. These are the remains of living organisms or the traces of such organisms (e.g. dinosaurs' footprints) which have been preserved in one way or another. Organisms that die in water and are rapidly buried in silt are commonly fossilised, hard parts being replaced or partly replaced by minerals, e.g. calcite, or else an impression may be made and retained in the earth itself. Animals and plants dying on land are much less commonly preserved because the natural processes of decay rapidly break them down to the original elements and nutrients out of which they were made.

Naturally occurring pools of crude oil in some parts of the world may preserve the bodies of creatures that fell into them. The complete bodies of mammoths have been found frozen into ice while twigs placed in certain silicous springs will be petrified within months. The exudate of trees may consist of a resinous material which hardens into amber and in this the whole bodies of insects may be preserved indefinitely.

Although fossils may be formed in a variety of ways it should be appreciated that the chances of an organism becoming fossilised are in fact very rare. It is thus not surprising that the fossil record is somewhat incomplete.

4. The time scale and main features of plant and animal successions. Early geologists had to work out the age of rocks from extrapolation of sedimentation rates but with the use of radio-active dating techniques based on the knowledge of the half life of certain isotopes much more accurate information is available. The main features of the geological succession, as revealed by fossils, are set out in Fig. 145.

5. The importance of zone fossils. Organisms that have remained virtually unchanged throughout geological time, such as the brachiopod *Lingula*, reveal nothing about evolution. Many organisms, however, are excellent zone fossils showing steady changes over very long periods of time (*see* Figs. 146 and 147).

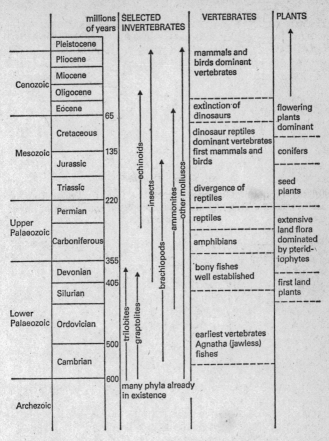

FIG. 145   *The geological time scale and major groups of organisms that flourished in the different eras.*

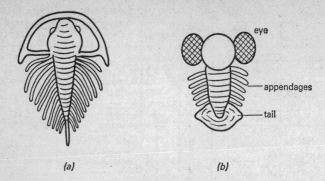

(a) (b)

FIG. 146 *Trilobites.*

(a) *Olenellus*, from the Lower Cambrian era. (b) *Cyclopyge*, from the Ordovician era. Trilobites are good examples of zone fossils of Lower Palaeozoic rocks.

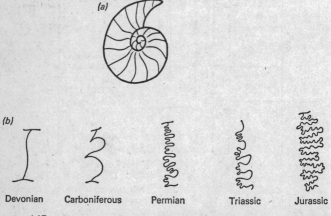

FIG. 147 *Ammonites, a type of fossil now almost extinct, are used as Mesozoic zone fossils.*

(a) Generalised form. (b) Evolution of complexity of suture pattern.

TABLE XXIV. IMPORTANT ZONE FOSSILS

Organism	Period
Echinoids (Sea urchins)	Cretaceous
Trilobites (Arthropods)	Cambrian → Mid-Devonian
Ammonites (Molluscs)	Upper Ordovician → Cretaceous
Graptolites (Protochordates)	Cambrian-Silurian

In such organisms evolutionary changes can be clearly followed, and rocks dated by examination of the zone organisms.

6. Link forms and living fossils. Belief in evolution implies that one phyla has given rise to another and therefore link forms which are of intermediate characters between phyla are of special interest. While such forms are found fossilised there are a number of cases of *living fossils* where link organisms have been found still alive in our contemporary flora and fauna.

TABLE XXV. IMPORTANT ANIMAL LINK FORMS

Phyla linked	Organism	Period
Agnatha-Gnathostoma	Lamprey	Holocene
Bony fishes-Amphibia	*Ichthyostega*	Carboniferous
	Coelocanth	Holocene
Amphibia-Reptilia	*Seymouria*	Permian
Reptilia-Aves	*Archeopteryx*	Jurassic
Reptilia-Mammalia	Ictidosaurs	Triassic-Jurassic
Annelida-Insecta	*Peripatus*	Holocene
Annelida-Mollusca	*Neopilina*	Holocene

7. General deductions on the course of evolution, derived from the study of palaeontology. Sufficient fossils are now available for quantitative studies to be made about the rate of evolution of selected species. A well-known example of this type of study was made by G. C. Simpson on the ancestry of the modern horse, *Equus*. He was able to show that the modern horse evolved from a small five-toed ancestor of the Eocene period and that it took sixty million years, eight genera, thirty species and fifteen million generations.

The idea that evolution proceeds in straight lines, called *orthogenesis*, is not confirmed by examination of the fossil evidence which rather supports the notion of continued selection acting on random mutations.

Some of the major groups of animals, for example the vertebrates, are thought to have evolved from larval forms of preceding groups. Such evolution is called *paedogenesis* and because it tends to involve soft-bodied larval forms it leaves no trace in the palaeontological record.

GEOGRAPHICAL EVIDENCE

8. Island faunas. Where species are isolated from the main bulk of the population, as for example on islands, rapid divergence may take place. Darwin was particularly impressed by the unique plant and animal species of the Galapagos Islands which are some six hundred miles off the coast of Ecuador in the Pacific.

In Table XXVI the numbers of unique species that occur in different phyla in the British Isles and the Galapagos Islands are contrasted.

TABLE XXVI.   UNIQUE SPECIES OF THE BRITISH ISLES AND
THE GALAPAGOS ISLANDS

| | Peculiar species | | | | |
	Land snails	Insects	Reptiles	Land birds	Mammals
British Isles	4	149	0	1	0
Galapagos Islands	15	35	10	30	0
	Non-peculiar species				
	Land snails	Insects	Reptiles	Land birds	Mammals
British Isles	83	12,551	13	130	40
Galapagos Islands	0	0	0	1	0

From this table it can be seen that the Galapagos Islands have a very high proportion of peculiar species, i.e. those that are not found elsewhere, while there are a very few species there which occur in other parts of the world. The British Isles, on the other hand, has very few unique species, and very many that occur elsewhere which reflects its comparatively recent separation from the Continent and the narrowness of the geographical barriers.

Among the many unique species of the Galapagos the diversity of the finches was particularly notable and Darwin related the various genera and species to each other and to a common ancestral form originating from South America (*see* Fig. 148).

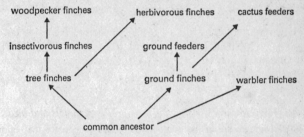

FIG. 148 *Diversity of finches in the Galapagos.*

Remote oceanic islands, such as the Galapagos group, have been aptly described as *laboratories of evolution* and their species, isolated from the main populations, may rapidly diverge. Even the islands around the Scottish coast and to the north each have their own sub-species of the wren, *Troglodytes*. There are sub-species of the wren in St Kilda, the Faeroes, Shetland Isles, Hebrides and Iceland and each varies in wing length and other particulars.

9. Major divisions of the earth's fauna. Wallace, the co-discoverer of the principle of evolution, made a survey of the general distribution of animal groups, in particular of the birds and mammals on the earth's surface. He showed that there are six major divisions (*see* Fig. 149) and that each of these has its own type-species and is separated from the others by great natural geographical barriers. One of these barriers is called *Wallace's line* and it is a trench of deep water that separates the flora and fauna of the Oriental and the Australasian groups.

The latter were cut off from the rest of the world in the Cretaceous period and are remarkable for the radiation of the marsupial mammals which, with few exceptions, are not found living elsewhere. There are many marsupial equivalents of the placental

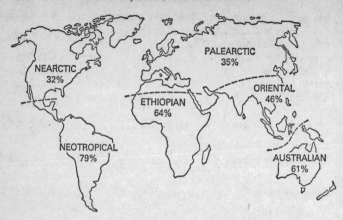

FIG. 149   *Natural divisions of the world's mammals and birds.*
The percentages are of the numbers of species endemic to the region.

forms such as mice and moles and wolves. These have all arisen from parallel adaptations to the demands of the environment.

There are literally hundreds of examples of speciation resulting from geographical isolation which have been studied and a very small selection of these is listed below:

(*a*) Insect species isolated by jungle on the upper reaches of mountains in Borneo.

(*b*) Salamander species of the Rockies and coastal districts of the North American continent.

(*c*) Snail species on the small islands of the Florida swamps.

(*d*) Great Tit species of Europe and Asia.

(*e*) Paeonies of the Mediterranean islands and Caucasus.

COMPARATIVE ANATOMY

Most facts of biology can be related to evolution in one form or another and there is a huge amount of evidence that evolution has occurred from a study of comparative anatomy.

10. Adaptive radiation. Underlying the divergence of species is often found the principle of *adaptive radiation* whereby a basic structure such as an appendage becomes modified for a variety of functions.

11. Radiation of appendages.

(*a*) *The pentadactyl limb.* This is the limb characteristic of tetrapod vertebrates. It has a single proximal bone, the humerus or femur, two distal bones, the radius and ulna or tibia and fibula, and a series of carpals and tarsals respectively followed by five phalanges or digits.

In all the tetrapod phyla the pentadactyl limb shows modification to a variety of habitats and the different types of locomotion involved. Thus to take the mammals we find the variations as shown in Fig. 150.

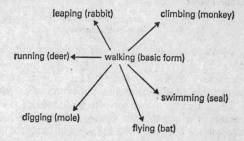

FIG. 150   *Modifications to the pentadactyl limb.*

(*b*) *Crustacean appendages.* While the lower crustacean, such as the brine shrimp, *Artemia*, have a number of similar appendages throughout the body the more advanced decapods such as the crayfish, *Astacus*, show variations of these homologous organs for a variety of functions (*see* Fig. 151).

(*c*) *Other examples of adaptive radiation.* It is an instructive biological exercise to list as many examples of adaptive radiation as one knows. An example that should be familiar to the student is the radiation of insect mouthparts from the basic chewing form seen in the cockroach to the elaborate piercing and sucking forms of the bee, mosquito, aphid and butterfly.

The specialisation of mammalian teeth to different diets is a further example (*see* VII, **5**).

From the plant kingdom the radiation of floral types from the supposedly basic ranalean flower provides a clear case of modification of homologous structures (*see* Fig. 152). Similarly the variation of leaves in mesophytes, xerophytes, climbing and insectivorous plants shows adaptive radiation for different environments.

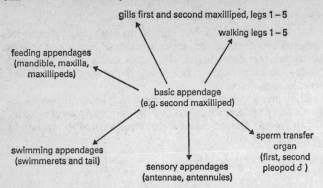

FIG. 151   *Specialised appendages found on the crayfish*, Astacus.

(*d*) *Evolutionary series.* The derivation of one anatomical arrangement from another is a clear indication of evolutionary affinities. There are many such series to be found in the study of the vertebrates of which the best known would be the heart and arterial arches (*see* Fig. 153), the kidney and its ducts, and the brain.

In plants the homologies of the life cycles through the bryophytes ⇌ pteridophytes → spermatophytes is a convincing demonstration of an evolutionary series.

(*e*) *Vestigial organs.* A further aspect of comparative anatomy is the presence of *vestigial organs*. These can best be explained by assuming gradual decrease in size and importance through change in functions and the selection of a gene complex that did not favour the development of the organ.

Examples are the tiny pelvic girdle of the whale and slow worm (a lizard) and the vestigial leaves of some xerophytes, e.g. *Ephedra*, and plant parasites such as dodder.

EVIDENCE FROM OTHER SOURCES

12. Embryology and evolution. It is found that the embryos of certain animals closely resemble each other although the adults are widely divergent. Such resemblance is taken as evidence of descent from a common ancestor. All vertebrate embryos are very similar in their early stages all, including man, going through a stage with gill slits and a tail (*see* Fig. 154). Similarly the

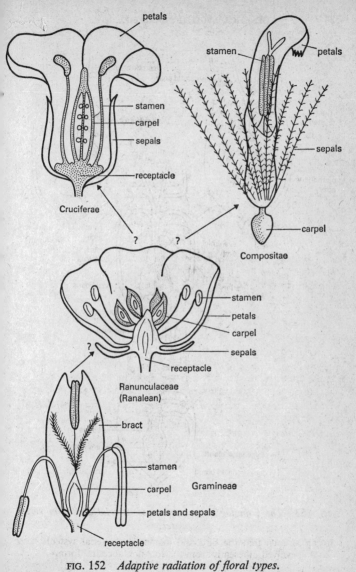

FIG. 152 *Adaptive radiation of floral types.*

It is likely that all flowers were derived from the primitive form of the Ranalean flower.

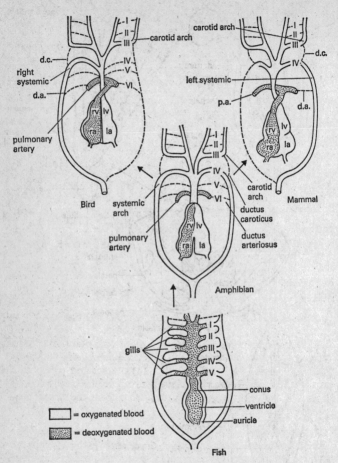

FIG. 153   *The homologies of hearts and arterial arches in verte-*
brates.

It can be seen that the bird and mammalian arterial systems have
evolved differently from the common ancestral form.

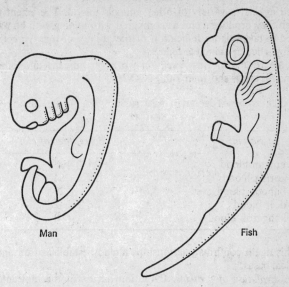

FIG. 154 *Very early embryos of a man and a fish showing the resemblance due to their evolutionary affinities.*

trochophore larva is common to the annelid and mollusc phyla and demonstrates their evolutionary affinities.

It seems likely that certain major evolutionary advances have taken place through a phenomenon called *paedogenesis*. This is where the larval form becomes sexually mature and produces a new adult organisation. Such may have been the case with the origin of the cephalochordates and fishes from the larval uro-chordate (sea squirts).

13. Physiological and biochemical evidence for evolution. The similarity of respiratory and other biochemical pathways in most living cells indicates the common ancestry of life. The distribution of such chemicals as arginine and creatine confirms the relationship of one phylum with another already deduced on other grounds.

(*a*) *Blood sera.* If the blood of a man is transferred to another animal such as a rabbit the latter builds up antibodies against it. These antibodies, contained in the rabbit's plasma, can now be tested against a variety of other animals' blood. The extent to which they produce a clot as compared with mixing with human blood is taken as an indication of the affinity of man with the other particular animal tested.

Some results of this *precipitin* work on humans and other mammals are as shown in Table XXVII.

TABLE XXVII.   PRECIPITIN REACTIONS WITH HUMAN BLOOD SERUM

Serum	Percentage of clot
human blood	100
gorilla blood	64
baboon blood	29
deer blood	7
kangaroo blood	0

This again confirms relationships already established on anatomical bases.

(b) *Evolution of proteins.* Evolution occurs at the molecular level and in recent years it has been possible to determine the sequence of amino-acids in protein chains. If the proteins of a living representative of an early form of life are compared with those of an organism known, on several other criteria, to be more advanced much information can be obtained about both the rates of molecular evolution and about the affinities of various organisms. Two examples are given.

(i) *Cytochrome c*, the ion containing H transfer enzyme of the mitochondria, has been analysed for a wide range of species.

From computer studies, coupled with other evidence, a picture of biochemical evolution and affinity for eight groups of selected organisms is indicated, as shown in Fig. 155. In this diagram the number of mutations away from the original form of cytochrome still found in the most primitive species is shown. (There is good reason to believe that each change in the position of an amino-acid in the peptide chain represents a mutation which has been perpetuated by natural selection.)

(ii) *Haemoglobin.* This is another protein that shows changes in the molecule. There is an approximate change of one amino-acid for every 10 million years. Haemoglobin in man consists of two α chains with 141 amino-acids each and two β chains of 146

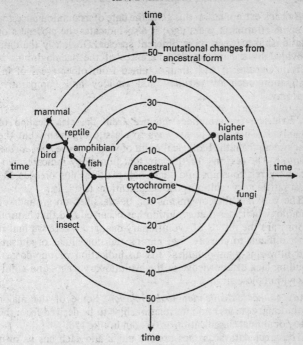

FIG. 155 *Biochemical evolution in cytochrome* c, *as found in a variety of organisms.*

amino-acids each. If a number of different mammals are taken and the differences in the number of amino-acids compared with man we get results such as shown in Table XXVIII.

TABLE XXVIII.   VARIATIONS IN HAEMOGLOBINS

Animal	Total differences in α chain
Horse	43
Pig	35
Cow	27
Rabbit	27
Gorilla	4

By back extrapolation this gives the date of a common ancestor at some 80 million years ago; it also indicates the closeness of relationship between these different species. Obviously the data provided by the molecular differences of the haemoglobins is fully in agreement with that obtained from other areas of investigation such as serology and geology and comparative anatomy.

14. Evidence for evolution obtained from the classification of organisms. There are many ways to classify organisms but the best system from the biological point of view is to use *homologous features*. The reasons for this are that homology indicates evolutionary relationships and that such relationships provide the most satisfactory multipurpose classifications (*see* I, 1).

The very fact that organisms can be classified on grounds of homology indicates that evolution has occurred and that natural groups are the results of evolutionary change. Where we find it most difficult to classify, e.g. among the unicellular organisms and flowering-plant families, it is an indication of our corresponding lack of knowledge of the evolutionary stages and affinities within these groups.

(*a*) *Animal classification and evolution.* Some of the major evolutionary stages and interrelationships to be derived from the study of animal classification are set out in Fig. 156.

The great natural groups are the phyla and each has its own basic characteristics. Within the phyla the various classes may show variations of these characteristics as a result of evolutionary adaptation.

There are very few animals that are difficult to classify and newly-discovered species tend to readily fall into one group or another.

(*b*) *Plant classification and evolution.* In many respects the classification of plants is less complete, at least on an evolutionary basis, than that of animals. It seems improbable that the origins and affinities of the higher plant families can ever be expressed satisfactorily and that some sort of artificiality in their classification must be accepted. There is controversy also about the major links between the groups.

An attempt to show the classification of plants in an evolutionary series is set out in Fig. 157. For details of the homologies of the life cycles reference should be made to III, 10 and X, 4.

It should be noted that the position of the bryophytes is

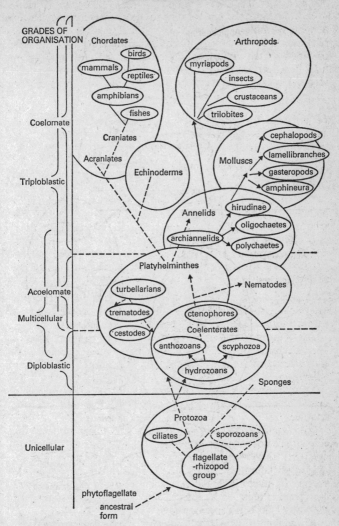

FIG. 156 *A classification of animals.*

The fact that such classifications can be built up on the basis of homologies of parts is strong evidence that evolution has occurred.

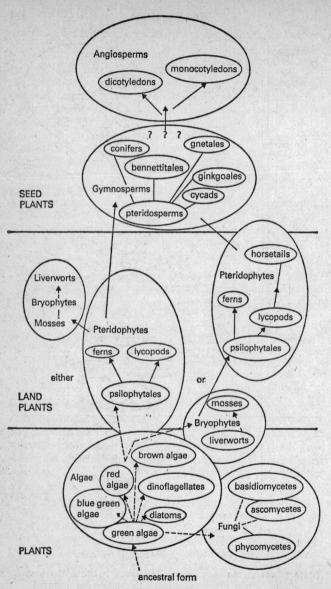

FIG. 157 *A classification of plants showing possible evolutionary affinities.*

uncertain, as is the origin of the pteridophytes and the spermatophytes.

15. Artificial selection and new species. The final evidence that
evolution has occurred is that deliberate human selection has over
the course of time caused many new types to emerge. If species
can be altered in this way by artificial selection there seems no
reason to suppose that selection in nature will not produce
speciation.

Species deliberately bred by man for his own ends include
sheep, cattle, horses, dogs, chickens, pigeons, budgerigars, apples,
roses, wheat and other cereals, potatoes and very many others.
It is probably true to say that very few domestic animals or crop
plants now exist in the wild form but all have undergone selection
for increased productivity, increased resistance to disease, or for
some other "desirable" end. In some cases extreme inbreeding
has produced biological monstrosities (e.g. hairless cats) which
would be most unlikely to survive in nature.

16. Summary on the occurrence of evolution. In the preceding
sections the evidence that evolution has occurred has been
briefly reviewed under its several categories.

These were as follows:

(a) Palaeontology.
(b) Geographical distribution.
(c) Comparative anatomy, embryology and physiology.
(d) Classification of plants and animals.
(e) Artificial selection.

It is clear that these are widely separated fields of biological
studies and all indicate quite definitely that evolution has taken
place. As stated at the beginning of this review the evidence presented is very largely circumstantial but nevertheless extremely
convincing. The alternatives to a belief that evolution has occurred seem to be quite untenable.

THE MECHANISM BY WHICH EVOLUTION
HAS OCCURRED

17. The Darwin–Wallace theory of natural selection. Darwin's
work *The Origin of Species by Natural Selection* was published
in 1859 and Wallace's paper *The Tendency of Species to Depart
Indefinitely from an Original Type* was published in 1855. They

are jointly credited with the discovery of the principle of natural selection. The principle is as follows:

(*a*) *Struggle for existence.* Many more offspring are born than survive; therefore competition must exist for survival.

(*b*) *Survival of the fittest.* The offspring of living organisms show variations from each other and their parents. Some of these variations are better than others; therefore individuals with favourable variations will tend to survive.

(*c*) *The origin of species.* Variations are passed on from parent to offspring; therefore the survivors of the struggle for existence will pass on their variations to their offspring. There will be a tendency to diverge away from the original type by the accumulation of favourable variations.

Darwin collected evidence for some of these contentions but had no real idea about the all-important source of variations and the means of their inheritance.

The theory of natural selection has been modified and extended especially by the work of the geneticists, and the composite theory of the mechanism of speciation is called *neo-Darwinism.* The evidence for, and the operation of, the process of neo-Darwinism will now be considered.

18. Competition for survival.

(*a*) *The number of offspring.* The numbers of offspring produced by different species may vary greatly but in all cases biological reproduction is a geometrical rather than an arithmetical increase, i.e. $2:4:8:16:32$ rather than $1:2:3:4:5$.

Some figures of the progeny (in terms of eggs laid or seeds produced) of a number of species are given in Table XXIX.

If we can assume that the population numbers of a given species do not fluctuate widely, every two individuals (or one for hermaphrodite species) will give rise to two surviving adults. It is therefore quite clear that the large majority of the offspring produced perish before they reach maturity.

Factors causing death of the offspring may include circumstances such as predation, lack of food, unfavourable environment, disease and many others. In all cases extreme competition for survival must operate at every stage of the life history.

(*b*) *Population explosions.* Very occasionally a situation occurs in nature where unlimited increase of a particular species may take place for a short period. Where such situations are found

TABLE XXIX.   NUMBERS OF PROGENY OF SELECTED ORGANISMS

Organism	Number per annum
Plants:	
Orchid	1,700,000
Radish	10,000
Mushroom	10^9
Animals:	
Oysters	16,000,000
Eel	10,000,000
Herring	30,000
Frog	1,000–2,000
Pheasant	14
Dog	4

NOTE: The hookworm produces 20,000 eggs per day: its life span is between five and seventeen years.

the very vast numbers that result indicate the severity of competition that normally controls the population. Well-known examples are the introduction of various species into the Australasian region. Rabbits and prickly pears built up vast populations in Australia, the latter colonising a region comparable in size to the British Isles. In New Zealand the bramble and the cabbage white similarly showed spectacular increases. These are cases where very highly adapted species are suddenly introduced into an environment in which there were no serious competitors, diseases or predators.

In parts of the Middle East swarms of locusts covering up to 1,000 km² have been encountered and special centres are set up to assess those combinations of environmental circumstances that lead to such *population explosions*.

Lemmings are also well known for the very large populations that can build up in a short period and for the migration from a central area; to a lesser extent the same phenomenon is seen in rat and mice populations.

That such explosive increases are rare indicates the normal balance of factors in nature, whereby competition for survival severely limits the numbers of any species.

(c) *The survival of larval plaice.* The North Sea plaice larva may be taken as an example of the type of factor which may determine survival. In this species the female lays some 20,000 eggs which hatch into small fishes with full yolk sacs. The food store lasts the fish only a short time and then it must fend for itself. It has been found that if the larval fish does not find a supply of the diatom *Coscinodiscus* within hours of finishing the yolk it will not survive.

Laboratory culture of the fish over this critical phase and subsequent release of the young fish into the sea should help to increase the population.

(*d*) *Sexual competition.* One aspect of the competition for survival that was stressed by Darwin is that of looking for a mate. In many animal species only a certain number of the adult population will actually produce the next generation.

This is particularly applicable to herd animals such as cattle, deer, seals, monkeys and others where a few dominant males sire all the progeny of the herd and the bachelor males do not. The fighting for dominance among the males of such species is a direct form of competition for biological survival.

VARIATION

19. The nature and sources of variation. Natural selection operates on the variations found in any population and the source and nature of variation will be described (*see* **20–22**).

20. Non-hereditable variation. These are the variations that can be observed in the phenotype due to environmental differences. Optimum nutrition for example produces larger organisms than poor nutrition and in some cases brings forward the onset of sexual maturation. The use of an organ or muscle may increase its size. In some plants leaves which are in the sun develop more layers of photosynthetic cells than those in the shade.

Such variations, although they may be adaptive to the particular environment in which the organism is living (i.e. help it to survive in such an environment), are not transmitted to the offspring because they are not due to changes in the genotype.

NOTE: Where an organism has been growing or living in a particular environment for several generations a specific gene complex may well have been selected, e.g. the prostrate form of grass on sea cliffs. Such forms are called *ecotypes*. Where such an organism is returned to a normal environment further generations may need to take place before genes whose action has been suppressed can manifest themselves or mutate back.

21. Hereditable variations or mutations. These are the variations that are of importance in the formation of new species. They may be due to gene mutation or to mutations causing changes in the chromosomes.

(*a*) *Gene mutations* may be brought about by α, β and γ radiation, by temperature, ultra-violet light and certain chemicals such as phenol. They involve chemical changes in the gene and affect the character the gene will determine.

Gene mutations are mainly harmful and upset the delicate balance of the genotype but occasionally favourable mutants occur which are the basis of evolutionary adaptation. Somatic mutation is not hereditable.

If a new mutation is favourable the rest of the gene complex tends to be selected in such a way that the new mutation is fully expressed. In other words the mutation tends to become increasingly dominant. If, on the other hand, the new mutation was of an unfavourable form then the gene complex would be selected to suppress its action, i.e. the mutant would become increasingly recessive.

It is no coincidence that so-called "normal" characters are mainly dominant and harmful characters recessive. A figure of 1 in 10^6 is given as the mutation rate under normal conditions but this is very variable and tends to be much higher in microorganisms. Back mutations returning towards the normal are known to occur.

(*b*) *Chromosome changes* may be structural where parts of chromosomes become inverted or removed from one place to another. Many of the mutations of *Oenothera* described by de Vries were of this type.

More important are mutations which involve a changing of the chromosome number, a phenomenon called *polyploidy*. Autopolyploids are individuals of whom the chromosome number has become some multiple of the haploid number other than 2. Thus triploids would have 3n chromosomes, tetraploids 4n chromosomes and so on. Polyploidy can be induced by the drug colchicine which interferes with spindle formation at cell divisions and prevents chromosome separation.

Allopolyploids are hybrids of two plant species that have undergone a doubling of the chromosome number which enables them

to reproduce sexually (*see* Fig. 158). Famous cases of allopolyploids are *Primula kewensis* and *Spartina townsendii*, both of
which have appeared in the last fifty years. Allopolyploidy is an

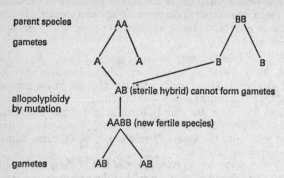

FIG. 158   *Formation of a fertile allopolyploid occurs by chance
doubling up of the chromosome number.*

important means of formation of new plant species, and as many
as half the known flowering plant species are thought to be poly-
ploids of various kinds.

22. Continuous and discontinuous variation. As we have seen
examination of the characters of a group of individuals within a
given population produces two distinct types of variation. The
first type is discontinuous variation where a certain character is
either present or absent, i.e. Mendel's peas were either tall or
short, wrinkled or round, yellow or green. Such characters are
determined by single genes and the inheritance of these follows
the pattern shown by Mendel.

The second type of character is one where continuous variation
exists and this applies to such characteristics as weight, height,
fertility, intelligence, etc. These characters are clearly due to the
combined operation of many genes and they do not give the
simple quantitative ratios of Mendelian inheritance. Continuous
variation tends to produce a normal distribution in the population and this range is subject to the same sort of selection that
applied to single-gene determined characters.

NATURAL SELECTION IN ACTION

While the theoretical mathematics of natural selection on populations has received a good deal of attention, data relating to actual field studies are still comparatively rare.

23. Evidence of natural selection. Darwin's own evidence on the occurrence of natural selection was almost entirely indirect, although he did carry out a few quantitative experiments on colonisation and competition in soils. Nevertheless, of all his contentions selection in action was the least well supported.

Since the publication of *Origin of Species* a great deal of work has been done on most aspects of evolution, but the experimental investigation of selection, especially in the wild, is still sparse. Such work is extremely difficult and it is not easy to be sure what factors are acting at any one time, besides which selective changes normally take a very long time.

In recent years microorganisms have been used to demonstrate population genetics and selection at work because they reproduce very fast and produce statistically significant numbers.

(*a*) *Selection in bacteria.* The colon bacteria, *E. coli*, has been subjected to the selective agent streptomycin and where this is added to a culture of the bacteria almost 100 per cent kill takes place. The few bacteria that survive represent mutant forms resistant to the antibiotic, and it is their progeny which replace the previously susceptible strain (*see* Fig. 159). Curiously enough such mutants do not survive well in a normal situation in competition with non-mutant bacteria; they are at an advantage only when the environment contains the antibiotic.

This process has unfortunately been taking place in wild populations of bacterial species since the introduction of antibiotics. Resistant strains of gonorrhea which can no longer be killed by penicillin have appeared, as well as many other pathogenic strains. For obvious reasons such resistant bacteria tend to turn up in hospitals.

(*b*) *Pest species and resistance to chemical control.* While the growth in resistance of pest species to the selective effects of chemicals were certainly not designed to show the process of natural selection, they certainly do demonstrate the process. Two examples are given.

(*i*) *DDT and insects.* The insecticide DDT first came into extensive use towards the end of the last war. It served a very

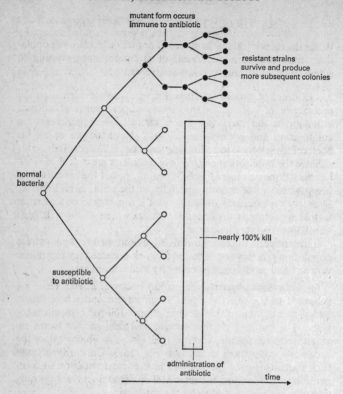

FIG. 159 *Selection in bacteria.*

useful role for the following two decades, although its use is largely discontinued these days. During the time it was in use it was noted that its effectiveness rapidly declined as, through natural selection, resistant mutants were replacing the previous susceptible populations.

(*ii*) *Warfarin and rodents.* The drug Warfarin was a successful means of controlling rats when first introduced in the late 1950s but within a decade resistant forms appeared (in Shropshire) and much larger doses were needed to kill such individuals. Of course the rate at which all these examples work depends on the generation time and fecundity of the individual organisms.

Thus bacteria change in response to selection pressures much faster than insects and these in turn a great deal more rapidly than mammals.

(c) *Industrial melanism*. Although associated with the peppered moth this phenomenon effects some 10 per cent of the 800 or so species of moth in this country, as well as being observed in various other types of insect. All the insects affected rest on trees or exposed surfaces during day and, where the surface has become blackened through industrial pollutants, natural selection favours the survival of melanic (dark) mutants and their progeny.

The phenomenon was first noted in Manchester in 1848 but has been not only observed in the British Isles but in other European countries, as well as industrial areas of the United States.

Large-scale surveys and experiments showed the black forms are favoured in industrial areas and also, in the British Isles, in

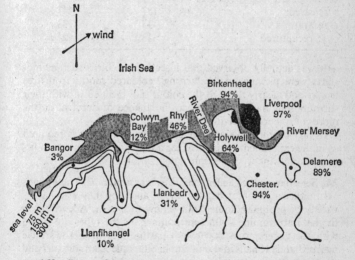

FIG. 160 *Percentage melanic (dark) peppered moths found in certain locations in north Wales.*

While representations of the percentage melanic moths are often related to the country as a whole, detailed surveys show the phenomenon also works on a smaller scale. The wind direction and the contours of the land indicated show that the areas to the west are virtually unpolluted and have a low percentage of melanic moths.

regions to the north-east of such areas, owing to the pattern of carbon deposit due to the prevailing south-westerly winds. Moths of one or the other colour were released in both clean and polluted areas and after a given period recapture was attempted. A typical set of results obtained by H. Kettlewell in the 1950s is given in Table XXX.

TABLE XXX. OCCURRENCE OF INDUSTRIAL MELANISM IN TWO AREAS OF THE U.K.

Area of release	White	Melanic
Dorset, 1955 (*unpolluted*)		
released	496	473
recaptured	62	30
% recaptured	12·5	6·3
Birmingham, 1953 (*polluted*)		
released	137	447
recaptured	18	123
% recaptured	13·1	27·5

It was actually observed that insectivorous birds such as redstarts were picking up the "wrong" coloured moths from tree trunks and clearly those of the colour that blended in with their surroundings had at least twice the chance of survival as the more conspicuous individuals.

In fact nothing like as large a selective advantage is required as in the experiments illustrated for a mutant of advantage to become rapidly incorporated into a population.

24. Selection on the banded snail, *Cephaea nemoralis*. These snails exist either as plain or banded forms, and they may also have a variety of ground colours from yellow to brown. According to the habitat examined, different numbers of the types are found with the general rule that rough pasture and hedgerows have mainly banded snails and uniform habitats such as short turf and the floor of beech woods have mainly unbanded snails. There is also a tendency for the ground colour to match the background of the vegetation.

The maintenance of these mixed populations was found to be largely due to selective predation by thrushes. Examination of the stone "anvils" on which these birds crack shells showed that banded shells were taken selectively from a uniform background

and that early in the year snails with a brown background were
not caught as often as in midsummer. Yellow-green snails escaped
predation in the green part of the year.

Cain and Shepherd found that natural selection favoured the
existence in the population of all types of shell banding and of
colours and, as can be seen from the data in Fig. 161, different

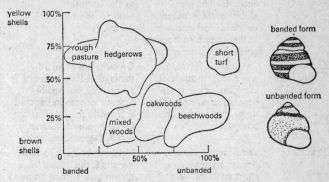

FIG. 161 *Percentage occurrence of banded and yellow-shelled
varieties of the snail* Cephaea nemoralis.

combinations of the colour and the bands produce best survival
in different environments. Thus uniformly brown shells are
inconspicuous in beechwoods against the uniform brown leaf
carpet, while yellow disruptively-banded ones are much better
camouflaged in rough pasture.

25. Other cases of polymorphism. There are very many examples
of polymorphism maintained by natural selection that are known
and only a few are given in Table XXXI.

26. The advantages of heterozygosity. It can be seen that it is of
advantage for a population to have a great deal of heterozygosity,
i.e. for genes to be represented by different alleles in different
individuals. Even where the phenotype shows the dominant gene,
which may be advantageous, it can be that a masked recessive
gene may, in new circumstances, become important for survival.
The environment changes continuously and there is no means of
knowing how it is going to change. If a population is homozygous
throughout, a sudden change will tend to bring about its complete extinction. If, on the other hand, a great deal of heterozy-

TABLE XXXI.   EXAMPLES OF POLYMORPHISM

Organism	Nature of polymorphic character	Advantage of the polymorphism
Man	Gene s causes, in hetero-zygotes, sickle-cell anaemia. The haemo-globin is abnormal and the red blood cells are sickle shaped. In homozy-gotes the ss gene is fatal.	Ss individuals are much more resistant to malaria, so the condition is of sur-vival value in tropical Africa.
Clover (and other plants)	Cyanogenic forms, C-, will release cyanide on crushing. There is an-other gene which speeds up or slows down this release.	C-plants are at an ad-vantage as they discourage herbivores, but spon-taneous release in the cold is harmful to the plant. C- gene more common in low well-grazed pastures.
Primrose	Heterostyly, whereby flowers have long or short styles or, very occasionally, medium-length styles.	Heterostyly leads to out-breeding, but where there are no insects then homo-styly and self-pollination advantageous.
Man	Blood groups.	Not yet understood.

gosity exists then new circumstances may be met by new combi-nations of genotypes previously unexpressed or even disadvan-tageous. It is not, for example, advantageous for a human to carry the gene for webbed fingers. If, however, the world became flooded so that man had to survive as an aquatic or semiaquatic animal the webbed digit gene might well have selective advan-tages! Obviously this is a far-fetched example but nevertheless it is true that it does seem to pay species of plants and animals to show a very considerable degree of genetic heterozygosity.

27. The Hardy–Weinberg equation. If we know the number of individuals in a given population who exhibit a double recessive character it is possible to calculate both the percentage of homo-zygote dominants as well as the percentage of heterozygotes. This can be done by a formula worked out by the two mathe-maticians Hardy and Weinberg in 1908. They reasoned as follows.

Let us suppose two genes exist in a population, say T and t, then the population will consist of individuals who are TT, Tt or tt. If we let the percentage of the gene T in the whole population is $p\%$ and the percentage of the gene t in the whole population is $q\%$ then clearly by definition

$$p+q = I \text{ (i.e. the whole population)}$$

If we now consider the frequency of the genotype TT it must be equal to the frequency of all the male gametes that contain T, that is p, multiplied by the frequency of all the female gametes that contain T, which is also p. Thus

$$\text{the frequency of genotype TT} = p \times p = p^2$$

By the same reasoning the frequency of the genotype tt must be equal to the number from the male gametes, that is q times the number from the female gametes carrying the gene, also q. Therefore the frequency of tt is $q \times q = q^2$

To obtain the frequency of the heterozygotes, Tt, we must take the frequency of the male gametes and the female gametes that can produce the heterozygote, which must be $p \times q = pq$. But it is also possible to produce a heterozygote if the different gametes come from either sex so the chance of the heterozygote is doubled to $2pq$.

Once again the total population of all the TTs, all the Tts and all the tts must equal the whole population, i.e. I. From this we get the equation

$$p^2 + 2pq + q^2 = I$$

With this relationship, as well as the knowledge that $p+q = I$, we can of course work out the values of TT and the importance of Tt, just as long as we start with the essential value for the homozygote tt. If a few calculations are made it can be seen how very large is the heterozygosity for most genes in a population and the importance of the condition appreciated. Let us take the Rhesus negative condition found in the blood of some 16 per cent of the British population; how many people will be carrying this recessive gene without showing it?

Rhesus negative $= rr = 16\%$

i.e. $\quad q^2 = 0.16$

thus $\quad q = 0.4$

as $\quad p+q = I$

$\quad\quad p = 1 - 0.4 = 0.6$

The number of individuals who carry the gene as heterozygotes, that is Rr, must be from the Hardy–Weinberg equation $2pq$

i.e. $2 \times 0.6 \times 0.4 = 0.48$

As a percentage that means nearly half the population! This fact may appear surprising but of course we do not have any real idea of the genetic advantage of the rhesus gene, either positive or negative.

The Hardy–Weinberg equation only works where there is random mating and no selection or mutation is taking place, but it is still a useful tool, especially in genetic counselling.

HOW SPECIATION OCCURS

We are now in a position to summarise the mechanism by which evolutionary divergence may be thought to occur.

28. The deme as a natural unit. It is clear that not all members of a species have the same chance of interbreeding but rather that any species is made up of small sub-units among which inter-breeding normally occurs. Such units are called *demes*.

29. Variation and selection within a deme. Within any given deme it is to be expected that continuous variations of multifactorial characters will exist and that in the course of time various muta-tions will occur. It is also likely that very few habitats are really identical in nature and because of this selection will operate differently on each deme so that some degree of difference of gene frequency will exist between one deme and another.

30. Relation of demes to the species as a whole. While the indi-viduals within a deme are more likely to breed with each other than they are with members of another deme, there will be a cer-tain amount of mixing of demes and interbreeding at their boundaries. The situation is comparable with the population of isolated country villages tending to be largely related to each other but to a lesser extent to families living outside the area.

The small amount of interbreeding and migration of indi-viduals between demes will have the effect of maintaining a com-mon pool of genes throughout the species such that all its mem-bers can breed together and produce fertile offspring.

31. Isolation and divergence. If a deme, or a group of demes, be-comes cut off either partially or completely from the main mass of the species, mutation and selection will cause *accumulation of* *new genes* and selection of genotypes that are not generally represented in the bulk of the species.

After prolonged isolation the genotype of such sub-units will be so substantially different from the common gene pool of the rest of the species that some degree of sterility will exist. By this is meant that a member of the original species, mated with an individual from a long isolated sub-unit, may hybridise successfully but when the supposed homologous chromosomes come together in germ-cell formation of the hybrid, incompatibility occurs. Viable gametes are thus not formed and the hybrid is sterile.

Eventually genetic divergence will become so great that even hybridisation is not possible and total breeding incompatibility between the original and the new species exists.

The stages of this process (*see* Fig. 162) can be seen in the use of these classificatory terms:

(*a*) *variety*—type within a deme;

(*b*) *race*—members of a partially isolated unit or units;

(*c*) *sub-species*—divergent from species by long isolation but still partially fertile with original species;

(*d*) *species*—a group of organisms that can breed together and produce fertile offspring.

The whole process is extremely slow and isolated demes or races where species divergence has begun are often reconnected by a breakdown in the isolating barrier so that a uniform gene pool is reconstituted. This seems to have been the case with our own species where initial divergence of the successful *H. sapiens* into Caucasoids, Negroids, Mongoloids, Amerindians, Australoids, etc. was stopped and the process largely reversed through interbreeding and migration of races. All members of the human race are potentially interfertile and it is biologically a single species.

PROGRESS TEST 14

1. How is biological evolution defined? **(1)**

2. Name three ways in which organisms may become fossilised. **(3)**

3. What were the main features of life in the Palaeozoic era? **(4)**

4. What are zone fossils? **(5)**

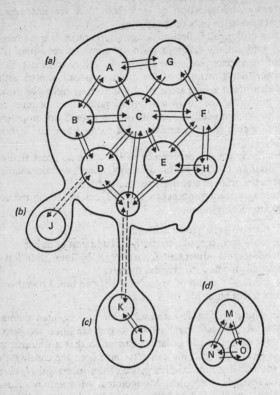

FIG. 162 *A scheme indicating the stages in speciation.*

(*a*) Interbreeding demes with sufficient genetic interchange to maintain reasonably similar genotypes. (*b*) Beginnings of isolation (geographical or otherwise), breeding with other demes less frequent, some divergence of genotype. (*c*) Almost complete isolation, genetic interchange now very rare, divergence of genotype causes partial sterility of hybrids. (*d*) Complete isolation in the population, genotypes M, N and O are substantially different from the original and hybrids can no longer be formed because of genetic incompatability. In the above model: A–I = original species; J = variety or race; K and L = sub-species; M, N and O = new species.

5. Which special features apply to island faunas and floras? **(8)**

6. Give four examples of isolating geographical barriers. **(9)**

7. In what way does the existence of adaptive radiation support the idea that evolution has occurred? **(10)**

8. Criticise the statement "man goes through a fish stage". **(12)**

9. What does "classification on the basis of homology" mean? **(14)**

10. Briefly state the Darwin–Wallace mechanism of evolutionary change. **(17)**

11. Why are population explosions rare in nature? **(18)**

12. What is meant by "mutation"? **(21)**

13. Why are most plant hybrids sterile but allopolyploids not? **(21)**

14. What is meant by "industrial melanism"? **(23)**

15. Why do not all members of a species have the same genotype? **(23, 24)**

16. How do species originate? **(28–31)**

EXAMINATION QUESTIONS

The figures in **bold** type indicate the marks allocated to each question or part question.

1. In evolutionary terms, what is meant by *adaptation*? **(4)** Illustrate your answer by reference to **three** well-defined examples (e.g. the fins of fish, the limbs of mammals). **(14)**

Cambridge, 1976

2. Give an account of the different sources of variation that occur in plants and animals. Illustrate your answer by reference to variations you have observed personally in wild species.

London, 1976

3. What do biologists mean by the following pairs of terms:
primitive/advanced;
undifferentiated/specialised;
analogous/homologous.
Illustrate your answer with appropriate examples.

Associated Examining Board, 1976

4. Although the evidence that evolution has taken place is substantial and accepted by a majority of biologists there are certain areas of ignorance and controversy. Review such areas in

which it appears to you that our knowledge is incomplete and unsatisfactory.

Associated Examining Board, 1976

5. Present a selection of the evidence which you think might convince a sceptical friend that evolution has taken place.

London, 1976

6. In industrial areas of England it has been reported that the dark form of a certain species of moth has increased in numbers compared with the lighter form. Explain how such a change may have come about. How would you attempt to demonstrate the truth of your explanation?

London, 1976

7. "Evolution is still occurring"—present evidence in favour of this view.

after Oxford and Cambridge, 1977

8. What are mutations? (4) List the types of mutations that occur and discuss three of them in detail. (16)

Oxford and Cambridge, 1976

9. In a large population 2 alleles B and b occur and the genotype frequencies are

BB	Bb	bb
49	42	9

Assume that mating is random and that all individuals produce roughly equal numbers of gametes and that genes B and b do not mutate. Show by calculation what the genotype frequencies are in the next two generations and comment on the rate of evolution of such a population. Discuss the effects upon this rate if (*a*) non-random mating occurs (*b*) mutation occurs (*c*) the population is small.

Oxford and Cambridge, 1973

CHAPTER XV

Ecology

1. The study of ecology. Ecology is not a laboratory study nor can it be understood from a textbook. To gather any useful information about ecological relationships in general or the specific autecology of an individual species, *data must be collected in the field*.

The aim of this HANDBOOK has been to provide the essential biological factual material required by the new "A" Level syllabuses. It is quite impossible to do this for ecological studies although these make up a part of all the new syllabuses.

In this chapter there will be set out the aims and methods of an ecological investigation together with a general consideration of the interrelationships existing between species. A possible approach to an autecological study will be given.

While all this information should help the student understand what ecology is about it must be pointed out that he will not actually know any ecology, or be qualified to answer questions about it, until he has carried out his own studies and made his own quantitative observations in the field.

THE BASIC IDEAS OF ECOLOGY

2. Definition. Ecology is the study of the living organism in its natural environment. It covers therefore all the factors that operate on an organism throughout its life, both the physical factors of the environment as well as the interrelationship of the organism with other living things. Nature is never stable and it is clear that the action of the whole environment on the organism and vice versa changes continually. Thus ecological studies are essentially dynamic.

Preliminary observations of ecological relationships may be descriptive, i.e. it may be recorded that herring stomachs contain large numbers of copepods, or that pale-coloured moths are more easily taken by birds than dark moths. Ultimately these observations have to be made quantitative to be useful, i.e. the *actual* numbers of copepods grazed in a certain time or the *exact* ratios between black and white moths consumed must be known.

A good deal of ecology is concerned with the collection of quantitative data and the analysis of this for significance or with the correlation between one observation and another. It is necessary to be able to use simple statistical methods to carry out such analyses.

3. Concept of the ecosystem. In order to reduce the very complex interactions that take place between organisms and their biological and physical environments, it is customary to try to divide up the latter into small well-defined units. Within such units, which are called *ecosystems*, the organisms have a close relationship with each other and with the physical problems imposed by the environment. To make clear what is meant, a list of typical ecosystems follows:

the sea shore,	the soil,	a jar of rain water,
hedgerows,	a tree,	a bird's nest,
a (beech)wood,	a cowpat,	a haystack,
a pond,	a dead animal,	a sand dune.

Larger areas which might still be considered as ecosystems are small islands, though where these are of any size so many varied systems are present that a single comprehensive survey is almost impossible.

It can be seen that the above ecosystems tend to provide a source of food and a suitable medium where plants or saprophytes and various types of animals can live. Some of the systems such as the sea shore and the soil have a permanent input of nutrient so they are repetitive and stable but others such as a cowpat or a corpse will change irreversibly with time (and so will its animal and plant populations).

4. The importance of ecology. Ecological studies impinge on many aspects of academic and applied biology. In particular they have thrown much light on animal behaviour and on natural selection in action. In the applications of biological knowledge to the conservation or destruction of a species ecological studies are essential preliminaries.

Some examples of applications of ecological studies under these headings is seen in Table XXXI.

The lists, especially of control studies, are endless. The opening

TABLE XXXI.   APPLICATIONS OF ECOLOGICAL STUDIES

Studies made for control of pests	Studies made for conservation and most efficient cropping
Wireworms	Fishes: Arctic cod, North Sea
Mosquitoes	plaice and herring, salmon,
Tsetse flies	trout.
Rabbits	Game: animals in South African
Rats	reserves.
Wheat bulb fly	Game: birds, pheasants, etc.
Colorado beetles	Forestry investigations.
Lice	Oyster fisheries, etc.
Prickly pears	
Cabbage butterflies	
Coypus	
Locusts, etc.	

up of new areas where the ecological balance is against man or his crops has stimulated vast amounts of ecological research. Sometimes these researches have not yielded the correct answers (as in the case of the addition of fertiliser to a Scottish loch to increase fish production, and in the settling of a ground-nut project in East Africa) and sometimes they have, as in the control of prickly pears in Australia or mosquitoes in Sardinia.

SAMPLING METHODS

5. Identification. The first step in an ecological investigation is to collect and identify the plants and animals that are present. Collecting methods must be appropriate for the size of the organism and for their identification. All sorts of specialised keys are available, keys such as *Soil Invertebrates, The Marine Diatoms, Insects of the British Woodlands, Sea-shore Fauna* and the *Concise Flora* are appropriate to a selected environment.

Identification keys are based on a system of questions being given several alternative answers for a particular anatomical feature of the unknown organism. Answering one question correctly leads on to another and so on until eventually the genus and species are determined. Such keys are familiar to botanists but they are equally applicable to zoological identifications.

In fact if one is dealing with an obscure ecosystem such as a decaying cowpat a specialist systematist may be necessary to identify the species present. This is not likely to be true of a rock pool or an oak wood.

6. Sampling. It is not usually possible to collect the whole population present in an ecosystem or to count it, nor indeed would it be desirable to do so. A technique of sub-sampling must therefore be used and we shall consider three ways of doing this for three different types of population.

(a) *Catch and release techniques* may be used for sampling populations of, for example, fishes or moths. A sample of the population to be investigated is first caught (by a trawl, or by a light trap in the case of the moths) and then marked in some suitable way (tags or rubber bands for fishes and fast-drying dopes for moths). The marked animals are then released and after a time a further sample of the population taken. It will be found that the number of marked individuals taken in the second sample bears the same relationship to the number originally marked as does the whole second sample to the whole population, i.e.

$$\frac{\text{number marked caught}}{\text{total numbers marked}} = \frac{\text{total number caught}}{\text{total population}}$$

An estimate of the total population can thus be made.

(b) *Random sampling* can be used for populations such as wireworms in fields. In the last war it became important for farmers to know the approximate numbers of wireworms present in their fields so they could sow a crop that would grow most efficiently. After considerable trial and error it was found that taking sampling positions in the field by use of random numbers some twenty samples per acre (60 per ha) provided an accurate assessment of the population when multiplied. It should be noted that there are a number of statistical "short cuts" by using double samples, etc. which allow approximate estimation of efficiency of sampling.

(c) *Quadrats and transects* may be made for organisms that remain fixed in one place such as barnacles and seaweeds and plants on land.

A *quadrat* is a square of known size which is placed over a randomly selected area of the site to be investigated and then the numbers of the species occurring in it are recorded. A number of samples may then be compared and by use of Chi^2 statistical tests, estimates of aggregation or dispersal made.

Transects are lines across the ecological site along the length of which the distributions of the various species is recorded. Such a method would be of use in investigating the zonation of seaweeds on a rocky shore or the succession of plants on the edge of a fen or swamp.

ECOSYSTEMS

7. Interrelationships within the ecosystems. It has been obvious from our knowledge of nutritional methods of organisms that food chains from the primary producer through herbivorous and carnivorous animals must exist. These relationships are examined more closely in ecological studies.

By detailed study of the organisms within the given habitat it is possible to construct food chains and food webs showing the nutritional interrelationships of the species. Examples of food webs for two ecosystems are given in Fig. 163. From the general pattern revealed in such food webs, a more detailed quantitative study may be made. As yet there are only a limited number of such studies available but certain generalisations may be made.

Thus it is clear that of the radiant energy falling on a square metre of vegetation only between one and five per cent is used in photosynthesis. Between the plants which are, of course, the primary producers in most ecosystems, and the herbivores, there is a reduction in biomass of something in the order of one hundred times. This is due to the matabolism taking place at each level of the system so that for every kilogram of food consumed only a small percentage is retained as increased body weight.

A further, though not so large, reduction operates between herbivores and the carnivores which tend to dominate the ecosystem. Figures obtained in the U.S.A. for grassland are as follows:

energy falling on a square metre per year $= 20 \times 10^8$ kJ
energy utilised by plants in same area $= 1.25$ per cent of
incident energy
productivity of plants $= 470$g dry weight per m^2
productivity of herbivores $= 0.6$g dry weight per m^2
productivity of carnivores $= 0.1$g dry weight per m^2

If we assume that stock is being reared and that the final consumer is man a further substantial reduction must occur.

It is thus quite clear that tapping the food chain lower down,

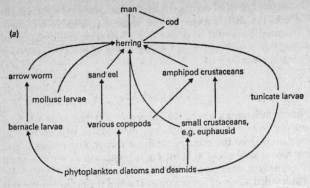

(a)

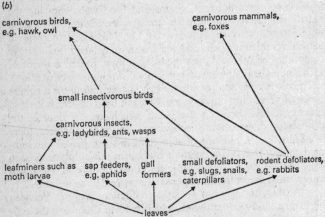

(b)

FIG. 163 *Food webs and chains* (a) *in the sea, for the herring,* (b) *in a wood for leaves and plants.*

It would be possible to add many more organisms to these schemes and similar sorts of schemes can be built up for any given ecosystem.

say at the level of the primary producer, will produce a far higher yield. As man is not able to digest the majority of plants this is seldom practical.

8. Models of production. Where sufficient data are available it is possible to derive equations which show the energy consumption and retention at each stage in the food chain.

If the data are sufficiently accurate such models can be used to estimate productivity and select the most efficient cropping rates of the population actually exploited by man. An example of such a model exists for the Georges Bank fishery of the North American coast and similar studies have been made for the North Sea cod and herring fisheries. The information these studies provide can be used to regulate the size of mesh used in trawling in an attempt to exploit the productivity chain with maximum efficiency.

SINGLE-SPECIES ECOLOGY

9. Autecology. As already described *autecology* is the study of the ecological relationships of a single species. While it is important to understand the general principles of ecology, more can be gained in a limited period from the detailed study of a single species than from a superficial survey of a whole ecosystem.

Examples familiar to the author are the autecology of the creeping buttercup, *Ranunculus repens*, and of the stinging nettle, *Urtica dioca*, and of the woodlouse, *Oniscus*. In commencing an autecological project it is advisable to set out a number of questions, the answers to which can be found with the apparatus available to the student.

Such questions might include the following:

(*a*) Which ecological niche does the species occupy?

(*b*) Why is it not found in other places?

(*c*) What are its chief competitors?

(*d*) What are its parasites and predators?

(*e*) On what does it live?

(*f*) What are its reproductive habits?

(*g*) How does its behaviour adapt it to its environment?

(*h*) To what extent will it survive in a foreign environment free from competition?

(*i*) How does it survive over winter?

(*j*) How is the species dispersed?

(*k*) What is the normal survival rate of the offspring?

Clearly some of these questions are only relevant to animal species and some to plants but if a reasonable collection of such questions are made and the work shared out between a number of groups in a class, a clear picture will emerge of the autecology of the particular species concerned.

It is easy in this type of project to discover something that is new. Some aspects of the organisms' relationships (e.g. water loss at different temperatures) are laboratory studies while other aspects can only be examined in the field. While the species chosen may have no economic importance the methods employed are those used by agricultural ecologists studying diseases, parasites, vectors and other problems of applied biology.

PROGRESS TEST 15

1. Define "ecology". **(2)**
2. What is meant by "an ecosystem"? **(3)**
3. Outline two methods of sampling the organisms present in a given environment. **(5, 6)**

EXAMINATION QUESTIONS

The figures in **bold** type indicate the marks allocated to each question or part question.

1. Explain with the help of a diagram what is meant by a *food web*. Why is the concept of a food web more realistic than that of a food chain? **(11)**

Plants are essential components of any food web. These harness the sun's energy in the production of organic compounds. Describe what happens to this harnessed energy in a food web. **(7)**

Cambridge, 1976

2. An ecosystem is a cycle of materials driven by a flow of solar energy. Explain and discuss this statement. **(25)**

Cambridge, 1976

Bibliography

GENERAL TEXTBOOKS FOR THE COURSE

Clegg, A. G. and Clegg, P. C., *Biology of the Mammals*, 4th edition, Heinemann Educational, 1975.

Freeman, W. H. and Bracegirdle, B., *An Atlas of Histology*, Heinemann Educational, 1966.

Grove, A. J. and Newell, G. E., *Animal Biology*, 9th edition, University Tutorial Press, 1974.

Marshall, P. T. and Hughes, G. M., *Physiology of Mammals and other Vertebrates*, Cambridge University Press, 1967.

Roberts, M. B. V., *Biology: A Functional Approach*, Nelson, 1974.

Simon, E. W. Dormer, K. J. and Hartshorne, J. N., *Lowson's Textbook of Botany*, 15th edition, University Tutorial Press, 1973.

Vines, A. E. and Rees, N., *Plant and Animal Biology*, 4th edition, Pitman, 1972.

Young, J. Z., *Life of Mammals: Their Anatomy and Physiology*, 2nd edition, Oxford University Press, 1975.

Young, J. Z., *Life of Vertebrates*, 2nd edition, Oxford University Press, 1975.

CLASSIFICATION OF PLANTS AND ANIMALS, BIOCHEMISTRY AND THE CELL (CHAPTERS I TO V)

Baldwin, E., *Dynamic Aspects of Biochemistry*, 5th edition, Cambridge University Press, 1967.

Barker, G. R., *Understanding the Chemistry of the Cell*, Edward Arnold, 1969.

Buchsbaum, R., *Animals without Backbones*, Penguin, 1971.

Dodge, J. D., *An Atlas of Biological Ultrastructure*, Edward Arnold, 1968.

Grimstone, A. V., *The Electron Miscroscope in Biology*, 2nd edition, Edward Arnold, 1976.

Harrison, K., *A Guidebook to Biochemistry*, 3rd edition, Cambridge University Press, 1971.

Heslop-Harrison, J., *New Concepts in Flowering Plant Taxonomy*, Heinemann Educational, 1953.

Heywood, V. H., *Plant Taxonomy*, Edward Arnold, 1967.

Hurry, S. W., *The Miscrostructure of Cells*, John Murray, 1965.

Phillips, D. C. and North, A. C. T., *Protein Structure*, Oxford University Press, 1973.

Ramsey, J., *The Experimental Basis of Modern Biology*, Cambridge University Press, 1969.

Shaw, A. C., Lazell, S. K. and Foster, G. N., *Photomicrographs of the Flowering Plant*, Longmans, 1965.

Shaw, A. C., Lazell, S. K. and Foster, G. N., *Photomicrographs of the Non-Flowering Plant*, Longmans, 1968.

Stace, C. A., *A Guide to Sub-Cellular Biology*, Longmans, 1971.

Stace, C. A., "Organelles", in *The Living Cell*, Scientific American, W. H. Freeman, 1965.

Swanson, C. P., *The Cell*, 3rd edition, Prentice-Hall, 1969.

Vickerman, K. and Cox, E. E. G., *The Protozoa*, John Murray, 1967.

PHYSIOLOGY (CHAPTERS VI TO XII)

Baron, M., *Organisation in Plants*, 2nd edition, Edward Arnold, 1967.

Bassham, "The Path of Carbon in Photosynthesis", in *The Living Cell*, Scientific American, W. H. Freeman, 1965.

Carthy, J. D., *The Study of Behaviour*, Edward Arnold, 1966.

Chapman, G., *The Body Fluids and their Functions*, Edward Arnold, 1967.

Fogg, G., *The Growth of Plants*, Penguin, 1970.

Fogg, G., *Photosynthesis*, 2nd edition, English Universities Press, 1972.

Friedmann, I., *The Mammalian Ear*, Oxford University Press, 1976.

Hardy, R. N., *Temperature and Animal Life*, Edward Arnold, 1972.

Hayashi, "How Cells Move", in *The Living Cell*, Scientific American, W. H. Freeman, 1965.

Hughes, G. M., *The Comparative Physiology of Vertebrate Respiration*, 2nd edition, Heinemann Educational, 1973.

Hughes, G. M., *The Vertebrate Lung*, Oxford University Press, 1973.

Jennings, J. B., *Feeding, Digestion and Assimilation in Animals*, Pergamon, 1965.

Kettlewell, "Facets of Genetics", in *Darwin's Missing Evidence*, Scientific American, W. H. Freeman, 1970.

McMinn, R. M. H., *The Human Gut*, Oxford University Press, 1974.

Manning, A., *An Introduction to Animal Behaviour*, 2nd edition, Edward Arnold, 1972.

Morton, *Guts*, Edward Arnold, 1967.

Neil, E., *The Mammalian Circulation*, Oxford University Press, 1975.

Ramsey, J., *A Physiological Approach to the Lower Animals*, 2nd edition, Cambridge University Press, 1968.

Randle, P. J. and Denton, R. M., *Hormones and Cell Metabolism*, Oxford University Press, 1974.

Richardson, M., *Translocation in Plants*, 2nd edition, Edward Arnold, 1975.

Rutter, A. J., *Transpiration*, Oxford University Press, 1972.

Schmidt-Knielsen, K., *Animal Physiology*, 3rd edition, Prentice-Hall, 1970.

Sutcliffe, J. F., *Plants and Water*, Edward Arnold, 1968.

Sutcliffe, J. F. and Baker, D. A., *Plants and Mineral Salts*, Edward Arnold, 1974.

Tata, J. R., *Metamorphosis*, Oxford University Press, 1973.

Tribe, M. A. and Whittaker, P. A., *Chloroplasts and Mitochondria*, Edward Arnold, 1972.

Wareing, P. F. and Phillips, I. D. J., *The Control of Growth and Differentiation in Plants*, Pergamon, 1970.

Weale, R. A., *The Vertebrate Eye*, Oxford University Press, 1974.

Whittingham, C. P., *Photosynthesis*, Oxford University Press, 1971.

Wilkie, D. R., *Muscle*, Edward Arnold, 1968.

Wilson, R. A., *An Introduction to Parasitology*, Edward Arnold, 1967.

Wooding, F. B. P., *Phloem*, Oxford University Press, 1971.

EVOLUTION, GENETICS AND ECOLOGY (CHAPTERS XIII TO XV)

Cloudsley Thompson, J. L., *Micro-ecology*, Edward Arnold, 1967.

Dowdeswell, W. H., *The Mechanism of Evolution*, 4th edition, Heinemann Educational, 1973.

Ford, E. B., *Evolution Studied by Observation and Experiment*, Oxford University Press, 1973.

George, W., *Elementary Genetics*, 2nd edition, Macmillan, 1965.

Phillipson, J., *Ecological Energetics*, Edward Arnold, 1966.

Watson, J. D., *The Molecular Biology of the Gene*, Benjamin, 1970.

White, M. J. D., *Chromosomes*, Chapman and Hall, 1973.

Examination technique

Most biology examination questions do not have quantitative answers and for this reason present certain problems both to the candidate and to the examiner. An attempt to examine biology more objectively is being made by a number of boards in the development of the *multiple-choice* type of question. In these the candidate is presented with a stem of information and a number of deductions that might be drawn from it and is asked to select the most valid item. While such questions have a good deal to recommend them (for example they can be marked mechanically) it does not seem to be the present policy of any board to use them for more than a part of the whole examination. It is therefore worth while to consider some ways in which conventional "essay style" questions should be tackled.

1. Mark schemes. An important point to remember is that the examiner will have in front of him a mark scheme and that this will consist of facts exactly related to the question set. It is a fair assumption that this scheme will add up to twenty in the normal theory paper.

The first thing that the candidate must learn to do is to weigh up the importance of the sections in a question that consists of several parts. Thus, to take some examples:

"What is meant by 'adaptive radiation'? Describe two examples of this phenomenon." (*Marking:* 6, 7, 7.)

"Discuss air and water as media for the uptake of oxygen and elimination of carbon dioxide. Describe briefly how oxygen reaches the muscles of a fish and a mammal." (*Marking:* 6, 6, 8.)

"What is a seed? Describe the mechanisms used to disperse seeds." (*Marking:* 8, 12.)

Having assessed the relative importance of the parts of the question, and this is not difficult to do with a little practice, the candidate must weight his own answer accordingly.

The second thing to remember is that the mark scheme is precisely tailored to the question set and that answers must be completely relevant and factual. Vague generalisations and long

361

rambling descriptions are useless. The mark scheme occupies only a few lines.

2. Allocation of time. Many candidates find the allocation of time under examination conditions very difficult, especially if they have been used to more or less unlimited time for essay writing. The majority of "A" Level biology papers are of three hours' duration so that only thirty-six minutes are available for any question (or thirty minutes in a two and a half hour paper). This means that a straightforward essay is not necessarily suitable.

Graphs, annotated diagrams, lists, sub-headings, flow diagrams are all ways in which a great deal of information can be put across in a short time. The underlining of key points is helpful to both examiner and candidate. Modern biological textbooks are not written in long uninterrupted sections nor is such a style suitable for examination answers.

A further point is that examination questions are not intended as vehicles for the candidate to show off irrelevant information. If a question on respiration states "No details of biochemistry are required" it means just this and therefore it is a waste of valuable time to include them in an answer.

3. The standard required. The same question, let us say on transpiration, might appear in an "O" Level and an "A" Level paper. Too often the former level of answer is submitted by a candidate taking Advanced papers. In the above example details of rates of water movement, xylem sizes, relative humidities, stomatal numbers, capillarity forces, etc., are all necessary for a good answer. Remember that physiology is a quantitative study of functioning and that purely descriptive accounts cannot be adequate.

4. Problem questions. Candidates are advised to choose problem questions which have definite answers or other questions where exact information is required. The more general type of question, e.g. "Describe the colonisation of land by living organisms" is very difficult to answer and may be better avoided, especially by mediocre candidates.

5. Practical examinations. As far as practical examinations are concerned several points are relevant. In the first place, many candidates fail to get down on paper the results of the time they have spent on a question. Remember that your script and perhaps dissection is all the examiner has and if you have not put the points across on paper about your technique, observations and conclusions they will not be credited.

A large number of people are not artistic but anybody can draw diagrams of biological material that are large, neat and accurate. Credit is given for these three things, especially the latter. It is quite useless to draw as your own dissection the half-remembered diagram from a book. No single dissection is the same as any other and your drawing should be a true picture of your own particular work.

6. Developing an examination technique. It is extremely import-ant that candidates preparing for examinations should answer questions under the same sort of conditions that they will actually face at the time of the examination.

A recommended method is to carry out some preliminary study of a topic, look at a question on it, draw up a rough plan of your answer, then, concisely and factually answer the question from the head, not the book, in the time allowed.

Index